I0033277

Keap Cookbook

Over 75 effective recipes for CRM optimization, marketing automation, and workflow mastery

Michelle Bell

‹packt›

Keap Cookbook

Copyright © 2024 Packt Publishing

All rights reserved. No part of this book may be reproduced, stored in a retrieval system, or transmitted in any form or by any means, without the prior written permission of the publisher, except in the case of brief quotations embedded in critical articles or reviews.

Every effort has been made in the preparation of this book to ensure the accuracy of the information presented. However, the information contained in this book is sold without warranty, either express or implied. Neither the author, nor Packt Publishing or its dealers and distributors, will be held liable for any damages caused or alleged to have been caused directly or indirectly by this book.

Packt Publishing has endeavored to provide trademark information about all of the companies and products mentioned in this book by the appropriate use of capitals. However, Packt Publishing cannot guarantee the accuracy of this information.

Group Product Manager: Aaron Tanna
Publishing Product Manager: Uzma Sheerin
Senior Content Development Editor: Rosal Colaco
Book Project Manager: Deeksha Thakkar
Technical Editor: Rajdeep Chakraborty
Copy Editor: Safis Editing
Indexer: Rekha Nair
Production Designer: Ponraj Dhandapani
DevRel Marketing Coordinator: Deepak Kumar and Mayank Singh

First published: June 2024

Production reference: 1200624

Published by Packt Publishing Ltd.

Grosvenor House
11 St Paul's Square
Birmingham
B3 1RB, UK

ISBN 978-1-80512-949-3

www.packtpub.com

In all things, be yourself; for you are uniquely divine, and that is a glorious thing to behold.

This is especially true in your marketing. Go boldly into the world and be your wonderfully authentic self, no matter what the gurus and experts say!

I promise the right people will find you.

- Michelle Bell

Foreword

Hey there, it's Lesley, the Product Owner of Automations at Keap. Let me tell you, I've seen firsthand how our software can transform businesses. In the small business world, finding the right tools can be like finding a hidden gem. And speaking of gems, Michelle Bell is a shining star in the Keap community. She's not your run-of-the-mill expert in online marketing and automation; think of Michelle more like your own personal problem-solving superhero, complete with her unmistakable vibrant orange hair and contagious personality. Michelle being ranked in the top 10 of the Keap Marketplace speaks volumes, but what truly sets her apart is her belief that every moment you spend working on your business should not only be profitable but also enjoyable!

Keap (formerly known as Infusionsoft) embodies an invaluable resource in the CRM landscape. Renowned for its robust suite of CRM capabilities, automation prowess, lead segmentation tools, and comprehensive business solutions, Keap has consistently adapted to meet the ever-changing demands of the market. Throughout its evolution, our reliance on our Certified Partner community has been pivotal. Their deep understanding of business needs and cutting-edge solutions ensures that we remain at the forefront of your industry, delivering unparalleled value and innovation.

Michelle's cookbook isn't just a collection of recipes; it's a key to unlocking the full potential of Keap's features. It's packed with strategies refined over years of hands-on experience, ready to streamline your processes and deepen your connections. Michelle's got this knack for turning complicated business jargon into practical, actionable strategies that even your pet goldfish could understand (if they were into that sort of thing).

Keap is all about empowering businesses to thrive, and Michelle's insights take it to the next level. By following her guidance, you'll be equipped to provide top-notch client support, automate tedious tasks, and boost your sales efficiency.

And hey, this isn't just about technical tips; it's about reclaiming your time and living your best life. Michelle's creativity knows no bounds, and her solutions are as effective as they are ingenious.

So, dive into this cookbook and elevate your business with Michelle's guidance. It's your chance to learn from the best and take your business to new heights. Enjoy the journey!

Lesley Oliver,

Product Owner of Automations at Keap

Contributors

About the author

Michelle Bell, the visionary CEO behind Virtual Work Wife, defies convention in the marketing landscape. Inspired by her own journey as a busy entrepreneur, Michelle founded *Virtual Work Wife* to empower others to reclaim control of their schedules, putting family first and work second.

Along with her business partner in Wifeboss LLC, she co-founded the Health Business Boss Institute, offering invaluable coaching, automation training, and strategic consulting to entrepreneurs and health professionals worldwide.

When she's not teaching, speaking, or on the road at dance competitions, Michelle enjoys time spent at home with her family. Through her journey, she exemplifies the transformative power of passion, perseverance, and gratitude.

About the reviewer

Heather Wells is the owner and CEO of Inspired Marketing Services Ltd., a small business marketing and automation company that she started over 15 years ago. As a single mother of 3, she started her business as a way to support her family while working from home. She has spent the last 15 years learning strategies, processes, technology, and principles that are key to the success of small businesses. Knowing how important family is, Heather's business is dedicated to helping other small business owners scale and grow their businesses using automation to help free up their time. She has worked with leaders in the digital marketing and online coaching space, helping them tame the chaos and grow their businesses to 6 and 7 figures.

Paul Sokol is a visionary American entrepreneur and automation virtuoso, renowned for his unparalleled expertise in leveraging Keap (formerly known as Infusionsoft) to revolutionize business operations. With an illustrious career spanning over two decades that includes writing the original "Infusionsoft Cookbook", he is recognized as a thought leader and trusted advisor, often sought after for his captivating keynote speeches, illuminating workshops, and insightful consulting services. His profound understanding of automated experience design, combined with his innate flair for innovation, has propelled him to the forefront of the industry, earning him widespread acclaim and admiration as a trailblazer in the realm of business automation.

I want to acknowledge Keap's co-founders Clate and Scott for having a big vision decades ago. The number of people's lives they have impacted cannot be counted. I also want to acknowledge Michelle Bell for not only being a great friend but for also carrying the torch on the automation cookbook. You rock girl!

Jade Olivia is the Owner and Founder of Jade Olivia Consulting and JadeOlivia.co, brands dedicated to empowering entrepreneurs. Since 2020, her consulting firm has offered community VIP consulting, automation strategy, and coaching. Under JadeOlivia.co, she launched Activate! for community-led engagement education and Timeless for mini-summits on marketing best practices. Previously, Jade was Strategic Partnerships Manager and Advisor Success Coach at Real Wealth Marketing, where she developed a marketing program for financial professionals.

Table of Contents

Preface   xiii

Part 1: Getting Started with Keap

1

System Overview   3

Technical requirements   3   Dashboard   11
Before we begin...   3   How to use it   11
Identifying your Keap version   5   How it works   12

Your application name   5   Setting up the Keap mobile app   12
Navigation   6   Initial setup   12
How to do it   6   Adding more than one account   13
How it works   11   Using Touch ID or Face ID for logging in   14

2

Designing Your Space   15

Technical requirements   16   Managing user accounts   21
Obtaining your Keap phone numbers   16   How to do it...   22
Getting ready   16   How it works...   24
How to do it...   17
How it works...   19   Setting your business profile   24
                                                 How to do it...   24
Creating your personal avatar   20   How it works...   26
How to do it...   20
How it works...   21   Connecting to your calendar   26
                                                 How to do it...   26

How it works…   28

Connecting to email   28

How to do it…   28

How it works…   29

Setting up appointment types   29

How to do it…   30

How it works…   32

Payment processing   33

Creating products   35

How to do it…   36

How it works…   37

Zapier integration   37

How to do it…   37

How it works…   37

There's more…   38

Google reviews   38

How to do it…   39

How it works…   39

3

Managing Contacts   41

Technical requirements   42   Importing contacts   50

Getting ready   42   Adding custom fields   52

Adding contacts manually   42   Introduction to tagging   56

Adding companies   45   Searching contacts   60

Grouping contacts   47

Part 2: Streamlining Communication

4

Communicating with Your Lists   65

Technical requirements   66   Getting ready   71

Email builder   66   How to do it…   71

Getting ready   66   How it works…   91

How to do it…   66   ## Text broadcasts   91

How it works…   70   ## Text templates   94

Choosing or creating   ## Broadcast reports   96
an email template   71

5

Managing Sales Pipeline   99

Creating your sales pipeline	**99**	**Generating invoices**	**111**
Getting ready	100	Getting ready	111
How to do it…	100	How to do it…	112
How it works…	104	How it works…	117
Creating quotes	**104**	**Creating checkout forms**	**117**
Getting ready	105	Getting ready	118
How to do it…	105	How to do it…	118
How it works…	111	How it works…	124

Part 3: Sales Pipeline Management

6

Marketing Forms and Landing Pages   127

Technical requirements	**128**	How to do it…	136
Public forms	**128**	How it works…	142
Getting Ready	128	**Landing pages**	**142**
How to do it …	129	Getting ready	142
How it works…	136	How to do it…	143
Internal forms	**136**	How it works…	158

Part 4: Automation and Reporting

7

Easy Automations   161

Technical requirements	**162**
Working with Easy Automations	**162**
How to do it…	162
How it works…	168

8

Advanced Automations   171

Technical requirements	171	Timers	181
Using the advanced automation builder	172	Communications	183
		Processes	183
How to do it…	173	**Working with decision diamonds**	**188**
Accessing automation version history	**176**	How to do it…	189
How to do it…	176	How it works…	192
Understanding goals	**178**	**Building an advanced automation**	**193**
Goals triggered by a contact	178	How to do it…	193
Goals triggered by a Keap user or automation	179	How it works…	199
Understanding sequences	**180**		

9

Reports   201

Technical requirements	201	**Working with contact tracker reports**	**208**
Working with sales reports	**202**	How to do it…	208
How to do it…	202	How it works…	211
How it works…	208		

Part 5: Integration and Optimization

10

Tying It All Together   215

Technical requirements	216	How to do it…	220
Managing documents and files	**216**	How it works…	223
How to do it…	216	**Manual automations**	**223**
How it works…	220	How to do it…	223
Posting manual payments	**220**	How it works…	226

Creating a deal manually   227
How to do it…   227
How it works…   230

Using notes   230
How to do it…   231
How it works…   232

Using tasks   232
How to do it…   233
How it works…   235

Requesting Google Reviews   235
How to do it…   235
How it works…   237

Managing your daily routine   237
How to do it…   238
How it works…   241

Utilizing the My day page   241
How to do it…   241
How it works…   243

11

Five Essential Automation Funnels   245

Technical requirements   246
Newsletter opt-in   246
How to do it…   246
How it works…   247

Contact Us   247
How to do it…   248
How it works…   249

Discovery call   249
How to do it…   249

How it works…   251

Lead magnet   251
How to do it…   252

Pay fail   253
How to do it…   253
How it works…   254

How essential sequences work
together   254

12

Data Management and Maintenance   255

Technical requirements   255
Merging duplicate contacts   256
How to do it…   256
How it works…   259

Managing opt-outs   259
How to do it…   259
How it works…   263

Managing bounced emails   263
How to do it…   263
How it works…   265

Fixing or deleting invalid emails   266
How to do it…   266
How it works…   267

Index   269

Other Books You May Enjoy   276

Preface

Learning any new system can be overwhelming. You've probably been asking yourself, "What are the first steps I should take when logging in to Keap? What is this platform supposed to be doing for me?" With over 20 years of marketing experience in an agency, I have often found myself having this conversation with new customers like you. It's the first question most people ask when we begin working together and that is what inspired me to write this book. The book will take a step-by-step approach to learning and optimizing Keap for your business. When you're done with this book, you'll be familiar with all of Keap's basics and even have a few automations in place.

Who this book is for

This book is for entrepreneurs, small business owners, and marketing professionals looking to harness the power of Keap to streamline their workflow, automate repetitive tasks, and supercharge their marketing efforts. Whether you're a seasoned expert or a newcomer to CRM technology, this book will guide you through the process of mastering Keap, regardless of your background or level of experience.

What this book covers

The book is broken down into 12 chapters; each designed to take specific steps to help you reach your overall goal of creating an automated central hub for running your business.

- *Chapter 1, System Overview*, helps you get acquainted with the Keap platform, understanding its core features and capabilities to lay the groundwork for effective implementation.

- *Chapter 2, Designing Your Space*, walks you through the process of setting up your Keap account, and configuring it to suit your business needs and preferences.

- *Chapter 3, Managing Contacts*, helps you learn strategies for efficiently organizing and managing your contact database within Keap to maximize engagement and conversion opportunities.

- *Chapter 4, Communicating with Your Lists*, explores best practices for creating and deploying targeted email campaigns that resonate with your audience and drive results.

- *Chapter 5, Managing Sales Pipelines*, helps you set up your sales process within Keap, from lead generation and conversion to effectively managing and nurturing leads through the pipeline.

- *Chapter 6, Marketing Forms and Landing Pages*, helps you create captivating lead capture forms and landing pages within Keap to expand your subscriber list and drive conversions.

- *Chapter 7, Easy Automations*, introduces easy-to-implement "when", "then", and "stop" logic to streamline your workflow and save time.

- *Chapter 8, Advanced Automations*, helps elevate your automation game with advanced strategies that use diverse conditional logic.

- *Chapter 9, Reports*, helps you harness the power of data within Keap to gain valuable insights into your marketing performance, enabling informed decision-making and optimization.

- *Chapter 10, Tying it All Together*, combines various components of Keap into a cohesive and effective system for running your business smoothly and efficiently.

- *Chapter 11, Five Essential Automation Funnels*, explores five essential automation funnels that can revolutionize your marketing efforts and drive results within Keap.

- *Chapter 12, Data Management and Maintenance*, helps you learn best practices for managing and maintaining your data within Keap to ensure accuracy, consistency, and compliance. *To get the most out of this book, there are three things that you'll want to do:*

    I.   This is an implementation-focused book, and you will get the most benefit if you do the steps as you follow along.

    II.   Remember, not all businesses are the same, so if you feel that an action or step does not align with your business process, set it aside.

    III. Test what you implement in real-time with valid phone numbers and emails. No matter how tempting it might be to skip a test or use a fake phone/email in your testing, there's no better proof that your automations works than getting an actual email in your inbox or text to your phone!

Finally, remember that completing a task is more important than achieving perfection. Anything you generate today can be improved upon tomorrow. Starting with something good and working towards greatness is perfectly acceptable.

Download the mobile app

Keap Pro/Max features a mobile app that lets users add contacts and companies on the go as well as access customer information.

	App Store for iPhones and iPads `https://itunes.apple.com/us/app/infusionsoft/id1421097870?ls=1&mt=8`
	Play Store for Android (only for 8.0 or later) `https://play.google.com/store/apps/details?id=com.infusionsoft.mobile.nimo`

Download the files

You can download the files for this book from GitHub at `https://github.com/ PacktPublishing/Keap-Cookbook`. In case there's an update, it will be updated on the existing GitHub repository.

We also have other code bundles from our rich catalog of books and videos available at `https:// github.com/PacktPublishing/`. Check them out!

Download the color images

We also provide a PDF file that has color images of the screenshots and diagrams used in this book. You can download it here: `https://packt.link/gbp/9781805129493`.

Conventions used

There are a number of text conventions used throughout this book.

`Code in text`: Indicates code words in text, database table names, folder names, filenames, file extensions, pathnames, dummy URLs, user input, and Twitter handles. Here is an example: "The "`From:`" address mirrors the user record you select in the dropdown menu."

Bold: Indicates a new term, an important word, or words that you see onscreen. For example, words in menus or dialog boxes appear in the text like this. Here is an example: "Click on the **Sales** card to open the side bar."

> **Tips or important notes**
> Appear like this.

Sections

In this book, you will find several headings that appear frequently (*Getting ready*, *How to do it...*, *How it works...*, *There's more...*, and *See also*).

To give clear instructions on how to complete a recipe, use these sections as follows:

Getting ready

This section tells you what to expect in the recipe and describes how to set up any software or any preliminary settings required for the recipe.

How to do it...

This section contains the steps required to follow the recipe.

How it works...

This section usually consists of a detailed explanation of what happened in the previous section.

There's more...

This section consists of additional information about the recipe in order to make you more knowledgeable about the recipe.

See also

This section provides helpful links to other useful information for the recipe.

Get in touch

Feedback from our readers is always welcome.

General feedback: If you have questions about any aspect of this book, mention the book title in the subject of your message and email us at customercare@packtpub.com.

Errata: Although we have taken every care to ensure the accuracy of our content, mistakes do happen. If you have found a mistake in this book, we would be grateful if you would report this to us. Please visit www.packtpub.com/support/errata, selecting your book, clicking on the Errata Submission Form link, and entering the details.

Piracy: If you come across any illegal copies of our works in any form on the Internet, we would be grateful if you would provide us with the location address or website name. Please contact us at copyright@packt.com with a link to the material.

If you are interested in becoming an author: If there is a topic that you have expertise in and you are interested in either writing or contributing to a book, please visit authors.packtpub.com.

Reviews

Please leave a review. Once you have read and used this book, why not leave a review on the site that you purchased it from? Potential readers can then see and use your unbiased opinion to make purchase decisions, we at Packt can understand what you think about our products, and our authors can see your feedback on their book. Thank you!

For more information about Packt, please visit packtpub.com.

Share Your Thoughts

Once you've read *Keap Cookbook*, we'd love to hear your thoughts! Scan the QR code below to go straight to the Amazon review page for this book and share your feedback.

https://packt.link/r/1-805-12949-X

Your review is important to us and the tech community and will help us make sure we're delivering excellent quality content.

Download a free PDF copy of this book

Thanks for purchasing this book!

Do you like to read on the go but are unable to carry your print books everywhere?

Is your eBook purchase not compatible with the device of your choice?

Don't worry, now with every Packt book you get a DRM-free PDF version of that book at no cost.

Read anywhere, any place, on any device. Search, copy, and paste code from your favorite technical books directly into your application.

The perks don't stop there, you can get exclusive access to discounts, newsletters, and great free content in your inbox daily

Follow these simple steps to get the benefits:

1. Scan the QR code or visit the link below

https://packt.link/free-ebook/9781805129493

2. Submit your proof of purchase
3. That's it! We'll send your free PDF and other benefits to your email directly

Part 1: Getting Started with Keap

In this part, you'll be introduced to the Keap platform, learn how to set up your account, and understand the foundational concepts essential for using Keap effectively.

- *Chapter 1, System Overview*
- *Chapter 2, Designing Your Space*
- *Chapter 3, Managing Contacts*

1

System Overview

Keap is a **CRM**, also known as **Contact Relationship Management** software. Basically, what that means is that it allows every person you come into contact with, whether they are a lead, buyer, affiliate, or something else, to be funneled into your CRM. By pulling every relationship into your hub you can then direct the flow of communication back to those people and track every eventual outcome.

The power of Keap's automation allows you to trigger when information should start or stop flowing, giving you the ability to create deeper conversations that are relevant and timely. Everything you do is logged in the contact record, giving you the documentation you need to manage employees or, should the need arise, to dispute chargebacks.

In this chapter, we will be making ourselves familiar with each area of Keap, how to navigate it, and what functions each area provides:

- Locating your unique ID and version of Keap
- Navigation
- Dashboard
- Mobile app

Technical requirements

For this chapter, the following are required:

- A Keap Pro/Max subscription
- An up-to-date Android or Apple phone for Keap mobile app usage (optional)

Before we begin...

It's important to note that there are three versions of Keap. While some information provided here is applicable to any version of Keap, **this cookbook is exclusively for learning to operate Keap Pro/Max.**

For reference, here are the three versions of Keap and how they differ:

- **Keap Pro** – This plan is great for businesses that are just starting out. It offers the following:

  A. Email marketing

  B. Sales, and marketing workflow automation

  C. Dedicated phone line (US and Canada)

  D. Automated lead capture and follow-up

  E. Lead and client management (CRM)

  F. Contact list and segmentation

  G. Google reviews

  H. Company records

  I. Email sync for Gmail and Outlook

  J. Task and note management

  K. Appointments:

    - Sales pipeline and analysis

    - Quotes, invoices, and payments

    - Landing pages and online sales

    - Text marketing (US and Canada)

- **Keap Max** is the next level up for businesses needing more customizable solutions. It offers the following:

  A. All the features of the Pro plan

  B. Lead scoring

  C. Lead source attribution

  D. Multi-page landing pages

  E. Upsells and discounts

  F. Promo codes

  G. Marketing analytics

  H. Advanced reporting

  I. Text marketing

- **Keap Ultimate** (formerly **Infusionsoft**) is the most robust and customizable of the three options. It offers a comprehensive array of features to aid businesses of all sizes in their growth:

  A. Advanced marketing and sales automation

  B. Custom user roles

  C. Email deliverability health reporting

  D. Shopping cart capabilities

  E. Referral partner management

  F. Fully customizable campaigns

  G. Powerful app integration

> **Important note**
>
> As of the time of writing, Keap Ultimate does not yet have Google reviews or dedicated phone-line capabilities.

One of the easiest ways to know what version you are using is by taking notice of where the navigation is placed.

Identifying your Keap version

Keap Pro and Max are built on the same foundation and therefore everything covered in this book applies to both versions.

Keap Ultimate has a unique foundation and while some content in this book may be helpful, most of the step-by-step instructions will not apply to Ultimate users.

Your application name

Each individual Keap account has its own app name, usually a series of letters and numbers. When you're contacting support, working with developers, connecting to external tools, or building an API, you are going to need to know your personal app name.

On the lower left-hand menu, you'll see a green circle with your initials or your user profile's avatar. Click to open it and your app name will be located just under your own name.

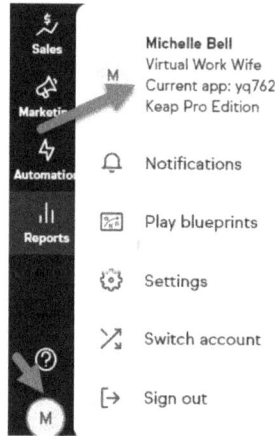

Figure 1.1 – Locating your app name

> **Important note**
>
> As a Keap user, you can upgrade or downgrade your version at any time. Please note that some functionality will be affected by this change. Make sure you've weighed all the options before proceeding.

Navigation

Let's walk through each section of the navigation, providing detailed insights into the purpose and functionality of key components.

How to do it

The navigation system in **Keap Pro/Max** is found in the upper-left corner of your screen and appears in a black box.

Clicking on the arrow just to the right of the **K** symbol expands the menu further.

Figure 1.2 – Identifying the Keap logo that opens the navigation

When you select a section of the navigation, the relevant section menu will open. Each section menu consists of two regions:

- Section menu (top)
- Related shortcuts (bottom)

Related shortcuts are items from other sections that are commonly used in connection with that area you have navigated to.

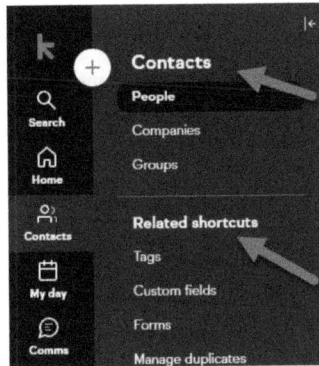

Figure 1.3 – identifying sections of the navigation menu

Here is a list of each section in the navigation:

- **Home**:

    A. **Dashboard** – Displays any reports or data widgets you have.

    B. **My Checklist** – Pre-written automation plays you can use to automate steps in your business. You can also create your own plays.

    We will cover customizing your **Dashboard** in more detail in a future chapter.

- **Contacts**:

    A. **Contacts** – Individual leads or customers you've added

    B. **Companies** – Allows you to view and store company-specific data as well as connect contacts associated with the company to give you a greater view of the organization

    C. **Groups** – Saved searches that you create based on common features such as tags and lead sources

    - **Related shortcuts**:

    A. **Tags** relate to contacts because we use them to create lists, a.k.a. groups, with a common attribute. Tags are also useful in automated workflows.

B. **Custom fields** are user-defined data fields that allow organizations to capture and store specific information that is not included in the standard set of Contact or Company fields provided by Keap.

C. **Forms** allows you to connect to your external form tools such as Typeform. You can also build your own custom public and internal forms within Keap. This is the same **Forms** section that can also be found under the **Marketing** section.

D. **Manage duplicates** is an essential tool for keeping your list healthy!

- **My day**:

  A. **Appointments** – Integrate your personal calendar and set up your appointment scheduling links

  B. **Tasks** – Assign yourself or others tasks as they relate to contacts moving through your automated systems

- **Communications**:

  A. **Business Line** – Keap Business Line, accessible in the United States and Canada, is a dedicated phone number for your organization and enables you to link your current mobile device and phone line to your Keap application.

  B. **Email broadcasts** – Send mass emails to specific subsets of contacts using data you've collected, tags, purchases, and other important segments of your contacts. This is the same **Email broadcasts** section that can be found under the **Marketing** section.

  C. **Text message broadcasts** – Similar to email, this lets you send mass texts based on your defined criteria. This is the same **Text message broadcasts** section that can also be found under the **Marketing** section.

- **Related shortcuts**:

  A. **Business profile** relates to communications as this profile contains the data people will see about you and your business.

  B. **Domains** allows you to connect your existing subdomains for landing pages and set up email authentication to improve deliverability. This is the same **Domains** section that can also be found under the **Marketing** section.

- **Sales**

  Sales is an essential area of your CRM. This is where you set up and manage your pipeline to track your potential deals, create your quotes, generate invoices, set up subscriptions, and build checkout forms so that you can automate collecting payments and repeat purchases:

  A. **Pipeline** – Manage various pipelines, each with its own set of unique stages.

  B. **Quotes** – Generate professional quotes and send them directly to your contacts, allowing them to easily click the **Accept Quote** button.

  C. **Invoices** – Send professional invoices and track payment status.

D. **Recurring payments** – Create subscriptions, memberships, and so on.

E. **Checkout forms** – Begin selling online quickly by taking payments from your website/ social media directly through your form. This is the same **Checkout forms** section that can also be found under the **Marketing** section.

F. **Products** – Add products to be sold via forms, quotes, and invoices.

G. **Promo codes** – Limit discounts by setting specific criteria and requiring a special code.

H. **Related shortcuts**:

- **Sales settings** relates to sales as this is where you set up your currency, payment processor, and sales tax. These are critical steps to take before you can create invoices and collect funds.

- **Marketing**:

A. **Forms** – Connect to your external form tools such as Typeform or build your own custom form within Keap

B. **Landing pages** – Create a custom page (or use one of the many templates) to capture leads

C. **Related shortcuts**:

- **Email broadcasts**

- **Text message broadcasts**

- **Checkout forms**

- **Domains**

- **Content Assistant** is an AI-driven feature that helps you write content for various assets based on your specific organization and target personas, such as lead magnets and consults

- **Automation**:

Keap comes with many templates as well as an advanced automation builder. Whether you're a beginner or an old pro, these automation tools help to bring the true CRM power to the table:

A. **Easy** – An easy-to-follow, step-by-step builder to get you started with automation

B. **Automation builder** – This is a more robust campaign builder that allows you to add more complex logic-based decisions on how a contact should proceed through your funnel

- **Reports**:

Keap provides many pre-built sales and contact tracking reports. You can also customize your reports, save versions, and create your own reports using the group feature in contacts:

A. **Sales reports**:

- **All sales** – Returns all sales with no grouping or totals

- **All sales (itemized) report** – Itemized report for each sale
- **Payments** – Payments grouped by month, day, and year
- **Receivables** – All payment plans that are due
- **Sales totals (by product)** – All sales grouped by product name
- **Failed invoice** – Payment plans that have failed

B. **Contact tracker reports**:

- **Tag tracker** – Lists all tags created
- **Web form tracker** – Details when contacts enter your system using your forms
- **Web form engagement tracker** – Catalogs how many new, unique, and repeat contacts you've collected via web forms
- **Sent email tracker** – Catalogs your sent emails with engagement stats for all of your contacts
- **Email batch results** – Displays information about all sent emails, including bounces
- **Email engagement tracker** – Manages contacts' engagement stats and confirmation status
- **Unsubscribe tracker** – Catalogs how many people have unsubscribed from the emails and any feedback they have submitted
- **Campaign enrollment tracker** – Displays unique contacts and which campaigns they are enrolled in
- **Campaign engagement tracker** – Tracks how contacts are engaging with your campaign content
- **Campaign progression tracker** – Shows contacts that are waiting for a step in a campaign sequence
- **Campaign sequence recipients** – Shows who has received an automation within campaign sequences
- **Campaign goal tracker** – Displays who has completed a goal in your campaigns

C. **?** (the question mark icon)

We all know "?" is the universal sign of help. Clicking on it opens the help menu on the right side of your screen. From here you can access live support and video training, and use the search feature to find relevant help articles.

How it works

Understanding the navigation is pivotal for you to leverage the full potential of Keap and enables you to manage contacts, execute sales and marketing strategies, implement automation, generate reports, and access support resources with confidence and proficiency.

Dashboard

Your Keap dashboard serves as a centralized hub that provides a visual representation of key metrics and data related to your business's interactions with its customers and leads. In essence, it is a snapshot of your day.

How to use it

Let's begin digging a little deeper into your CRM. When you first open Keap, you will land on your dashboard. This offers a snapshot of what's going on in your business:

- The dashboard shows you how many contacts you have both in total and also broken down into various date ranges
- It displays relevant sales data including repeat clients, sales, and pending quotes
- You can manage your daily activities such as tasks and appointments here too

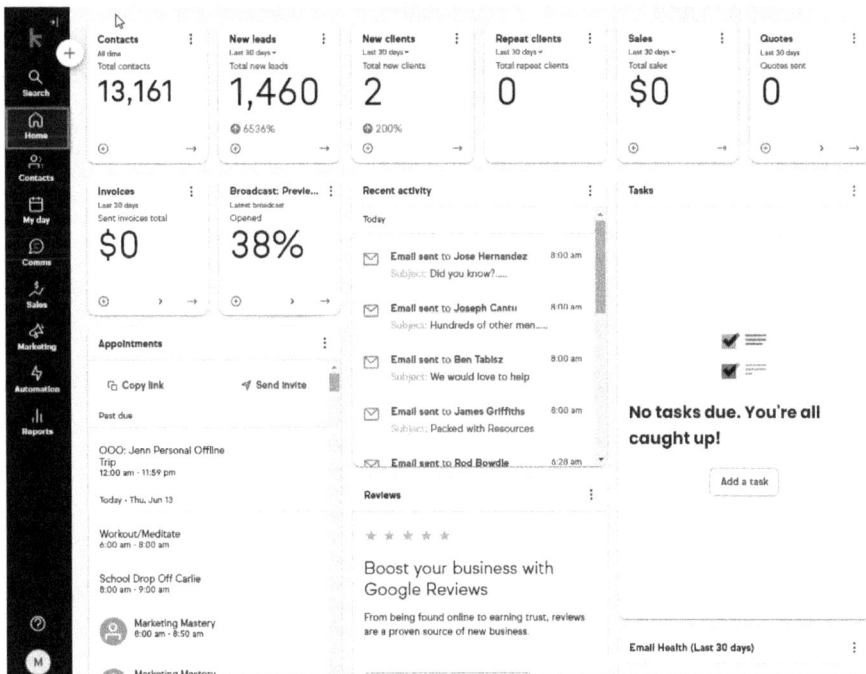

Figure 1.4 – Dashboard overview

How it works

By incorporating your Keap dashboard into your daily routine, you create a centralized command center that empowers you to stay organized, make informed decisions, and proactively manage your tasks, funnels, leads, and cash flow. This daily practice enhances efficiency, promotes accountability, and contributes to the overall success of your business.

Moreover, Keap's customizable dashboard allows you to tailor your workspace to your specific needs. With the ability to toggle various widgets on and off, as well as rearrange them according to priority, you can fine-tune your dashboard to focus on what matters most at any given time. This flexibility ensures that you maintain a laser-like focus on key metrics, tasks, or objectives, further enhancing your productivity and streamlining your workflow.

Setting up the Keap mobile app

Keap Pro/Max comes with a mobile app that lets you add contacts and companies on the go as well as access customer info. You can easily manage tasks and appointments and apply notes and tags to contacts, ensuring they move through your automations seamlessly.

Use the app to receive mobile reminders and alerts so you never miss important dates and activities. The Keap mobile app ensures you and your customers have a winning experience.

You can download the Keap app from the App Store for iPhones and iPads: `https://itunes.apple.com/us/app/infusionsoft/id1421097870?ls=1andmt=8`

Download the app from Play Store for Android (only for 8.0 or later): `https://play.google.com/store/apps/details?id=com.infusionsoft.mobile.nimo`

Initial setup

Once you've successfully downloaded the app, it's time to open it and log in:

1. Enter the same email address you use to log in to the website.
2. Complete the security check, enter your password, and tap the **Log in** button.
3. If you have been working primarily from your inbox, allow Keap to import your contacts.

Figure 1.5 – Mobile app login screen

Adding more than one account

If you have more than one account associated with Keap, you can access all of them after you log in:

1. Enter your username and password.

2. Tap the **Log in** button and you will reach a screen showing all of your available apps.

3. Scroll through and select an app. You have access to your business line, contacts, and tasks and can update your settings.

For a more expansive tutorial on the app, click the question mark icon in the top-right corner of the home screen.

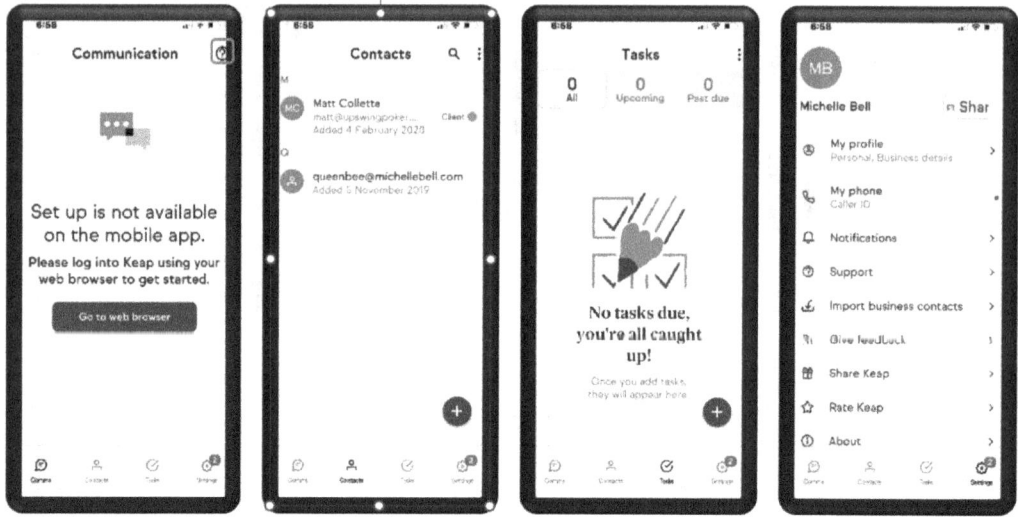

Figure 1.6 – Customizing your mobile app

Using Touch ID or Face ID for logging in

Users with an iOS device that supports Touch ID or Face ID can enable this feature for future logins:

1. Enter your username and password.
2. Tap the **Log in** button.
3. The second time you are prompted to log in, the app will ask permission to use Touch ID or Face ID.
4. Tap **OK** to grant permission and follow the on-screen instructions.

2

Designing Your Space

Setting up your branding and basic information before implementing outward-facing marketing is crucial to establishing a strong foundation for your **customer relationship management** (CRM) system. By defining your brand identity, including logos, colors, and style, you ensure consistent and recognizable communication across all lead and/or customer touchpoints.

Accurate and comprehensive basic information about your business, such as contact details and product/service descriptions, will enhance the overall customer experience, which creates trust, credibility, and brand loyalty right from the start.

Maintaining a cohesive customer experience across all touchpoints also reinforces the brand's message and its unique selling points. Ultimately, consistent branding not only drives higher email open and click-through rates but also strengthens the overall brand equity and customer relationships, leading to long-term business success.

Additionally, the overall usage of your CRM system will be more efficient if you take the time to create consistent branding before you begin marketing. You're going to want to have all the assets in place before you begin creating complex email marketing or workflow campaigns!

We'll cover the following recipes in this chapter:

- Obtaining your Keap phone number
- Creating your personal avatar
- Creating new user accounts
- Setting your business profile
- Connecting to your calendar
- Connecting to email
- Setting up appointment types
- Payment processing

- Creating products
- Zapier integration
- Google reviews

Technical requirements

For this chapter, the following are required:

- Your logo
- Access to your preferred calendar tool
- Your headshot
- Your preferred email address
- Access to your chosen payment processing account
- Access to your Google Business account
- Zapier (optional)

Obtaining your Keap phone numbers

Keap is an excellent tool for helping you improve communications with your contacts. As a small business, one of the most frustrating things you will deal with is separating your personal space from your business enterprise. Where we see this most often is in the tools we use to communicate. Using your personal cell phone to conduct business means always being "on the clock." Conversely, who wants to carry multiple devices?

Why do you need a business line and a marketing number?

- Your Keap marketing number is used for automated text message broadcasts to many contacts at once
- Your Keap business profile is for one-to-one communications and small-group texts
- In the event that a contact opts out of your Keap marketing number, you can still use your Keap business line to contact them

Getting ready

If you haven't already downloaded the mobile app, which we covered in *Chapter 1*, you will need to do so before proceeding. Then, make sure you are logged into the Keap website.

How to do it...

In this recipe, we will walk you through the steps to create your free business numbers, which you will access via the Keap mobile app and the website:

Phone numbers for a business line

1. Click on the **COMMS** tab in the left-side navigation bar to open the menu and choose **Business Line**.

2. Click the **Get my phone number** link.

3. Keap will now ask you to enter your area code in order to find a number that aligns with your location:

Figure 2.1 – Choosing your area code

4. Enter your preferred area code in the box, as shown in *Figure 2.1*.

5. Next, you will need to agree to the terms and send a code to your mobile number to connect your business line to your mobile device.

Selecting a marketing number

1. Begin by selecting **COMMS** from the navigation bar and clicking on **TEXT MESSAGE BROADCASTS**.

2. You will have the option to watch a video or to skip the tour. Choose **SKIP THE TOUR** to continue.

3. On the next screen, begin by adding your business name in the provided box and clicking the **NEXT** button.

4. Next, you will need to check all the boxes, agreeing to the terms, conditions, and privacy policies:

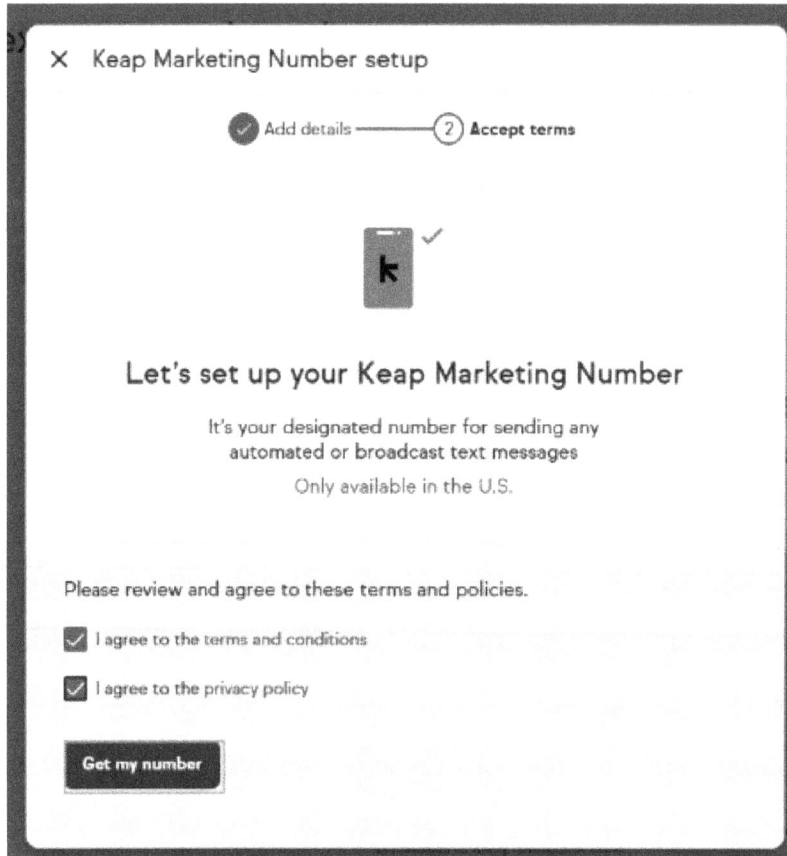

Figure 2.2 – Agreeing to the terms

5. Complete this step by clicking the **Get my number** button.

6. In the drop-down box provided, you will need to declare the intention for your texting (i.e., customer care, marketing promotion, etc.).

7. Carefully review your answers and then click the **SUBMIT** button to proceed. Your new Keap number will be established! Then, click on the **Continue** button:

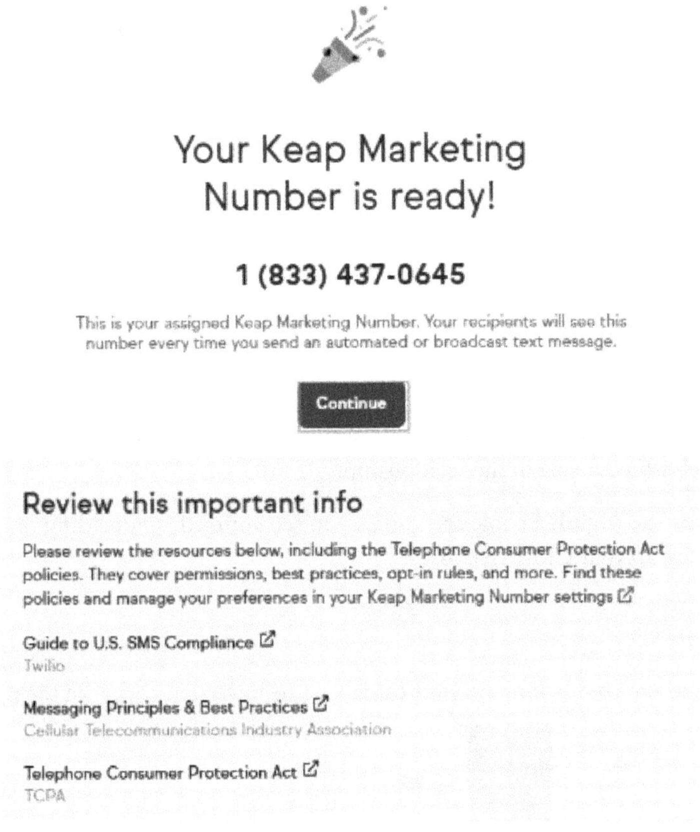

Your Keap Marketing
Number is ready!

1 (833) 437-0645

This is your assigned Keap Marketing Number. Your recipients will see this
number every time you send an automated or broadcast text message.

Continue

Review this important info

Please review the resources below, including the Telephone Consumer Protection Act
policies. They cover permissions, best practices, opt-in rules, and more. Find these
policies and manage your preferences in your Keap Marketing Number settings ☑

Guide to U.S. SMS Compliance ☑
Twilio

Messaging Principles & Best Practices ☑
Cellular Telecommunications Industry Association

Telephone Consumer Protection Act ☑
TCPA

Figure 2.3 – Confirming your number and reviewing the guidelines

How it works...

Keap searches a repository of available phone numbers and gives us a few recommendations to choose from.

Don't waste time looking for a "perfect" or easy-to-remember number. No one memorizes numbers anymore! We just store them in our phone contact list, usually, with quirky nicknames so that we recall who they are.

Text messaging compliance is regulated; therefore, it is necessary to provide details about how people will opt in to your list and how often you will text. Additionally, you will need to verify your company name, website, phone number, and address.

As we discussed in *Chapter 1*, you can view contacts and tasks from your mobile app. Now, you can send texts and make calls from your business line via your app or the website as well.

This gives you even more power to stay on top of leads and clients and ensures better follow-up. For example, when a call has ended, the Keap mobile app will prompt you to create a note about what happened on the call, as well as a task to follow up with later.

During your "off" hours, you can also use the "do not disturb" feature, which will automatically respond to incoming calls, business line calls, and texts with a message that says you are unavailable.

> **Important note**
> You can call or text any US or Canadian phone number with your Keap business line.

Creating your personal avatar

Your personal avatar or profile is what people will see when receiving emails from you. This is your opportunity to establish your role in the relationship.

Are you the CEO or the yogini? People see you as you see yourself, so deciding on a persona is a critical step in setting up your CRM system.

How to do it...

1. We're going to start by clicking on your initials in the lower-left corner of the navigation:

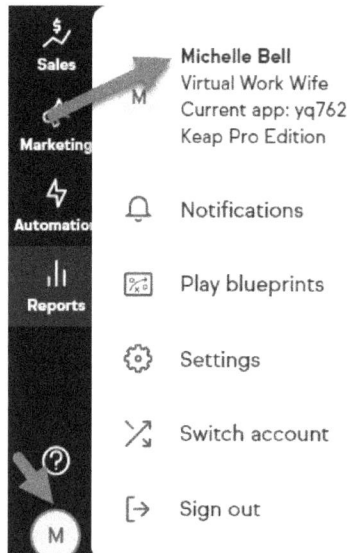

Figure 2.4 – Navigation to profile

2. Next, click on your name in the pop-out menu. This will bring you to your profile.

> **Important note**
>
> The email address that you log in with will initially be in your profile. However, that may not be the best email for you to use when sending business communication. For example, you might have set up your login using a personal Gmail address, but you want your emails to be sent out from a professional domain. So, the first thing you're going to do is verify that the email address in your profile is the one you want the world to see as your business address.

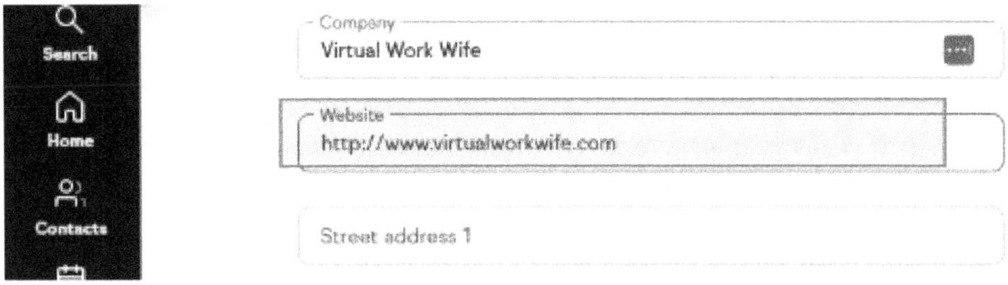

Figure 2.5 – Add your website address

3. Add your website address, making sure to use `http://` or `https://` to begin your URL.

4. Continue to update your profile by utilizing the space provided and then click the **UPDATE** button to save and preview your email signature changes.

How it works...

Your personal avatar is the face you present to the world in your business communication. It's a representation of your role and identity in the professional realm. By carefully configuring your profile and utilizing a suitable email address, you can make a strong and lasting impression on the recipients of your emails. Your personal avatar is a valuable tool in establishing and nurturing your relationships with clients, customers, and associates.

Managing user accounts

As new employees enter your business, they will likely need access to your CRM system. Conversely, you may also find yourself needing to remove or deactivate users when an employee leaves your business. While the process of adding a new user is fairly simple, removing a user has the added layers of reassigning contacts, deals, and tasks to another user so that your leads and customers continue to be managed.

Once your new employee is trained, you may find yourself needing to reassign those tasks, contacts, and deals again to redistribute the workload among your team.

In this recipe, we will go over the steps for creating/deactivating users in Keap.

How to do it...

Adding additional users to your CRM system by defining them as job roles rather than personal identities can have massive benefits in terms of growing your business.

In general, people will come and go from your company, but the job roles will remain the same. Rather than creating a new profile every time you have a new employee, consider creating a job role and then simply updating the password when a user leaves.

To do this, you would first start by creating your email address for that job role – for example, `support@`, `sales@`, or `advisor@yourwebsite.com`; these are all job roles within my company.

Use these addresses to create your login to Keap and any other tools the role will need. Now, when you have a change of staff, you only have to change the passwords rather than deactivating personal accounts, creating new ones, reassigning tasks, and updating templated content.

Creating job roles will save you time and money in the long run.

Adding users in Keap

1. We're going to start by clicking on your initials in the lower-left corner of the navigation.
2. Next, click on **SETTINGS** in the pop-out menu. This will bring you to the **Settings** page.
3. Select **USERS** from the menu.
4. Click the **ADD A USER** button in the top-right corner.
5. Enter the first name and email address for the person or job role that you want to add to your CRM system and choose the appropriate permission level. There are four permission levels: **Admin**, **Limited Admin**, **Manager**, and **Staff**:

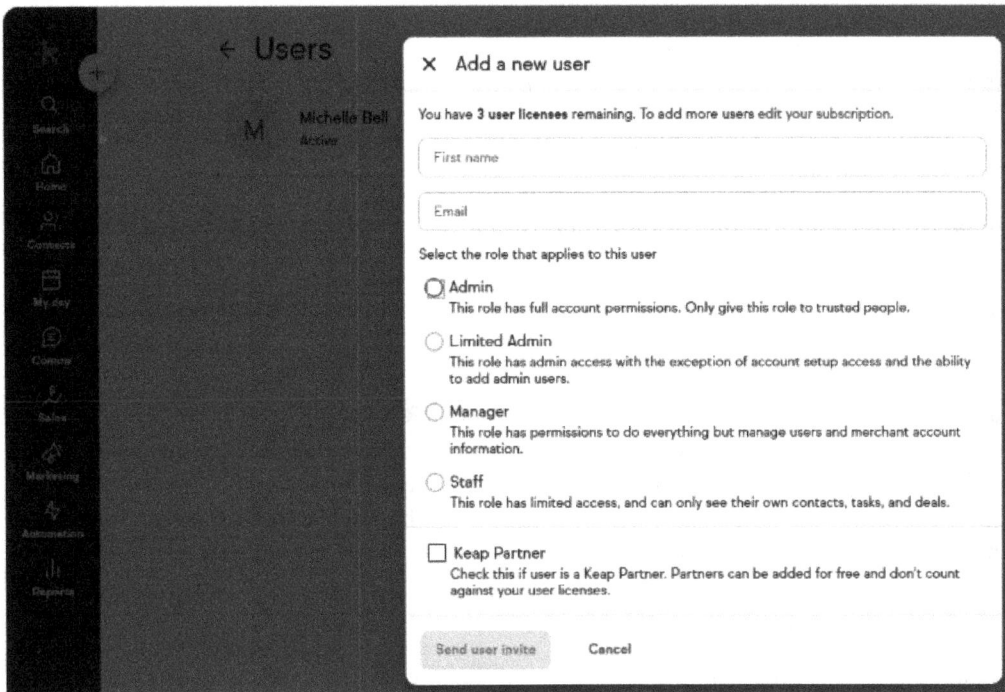

Figure 2.6 – Identifying the user type

> **Important note**
>
> If you are adding a Keap Partner, you must check the **Keap Partner** box to ensure you do not get charged for the user. Keap Partners can be added for free. In most cases, they will need admin-level access to support you.

6.   Click the **Send user invite** button to invite your team members to your CRM system.

You will not be able to send broadcast or individual emails from, or assign tasks to, this user until they have accepted the invitation to create their login profile.

Deactivating users in Keap

1.   Follow *steps 1–3* from the *Adding users in Keap* section.

2.   Next, click on **Deactivate User**.

3.   In the pop-out, confirm your action by clicking **Deactivate user**.

How it works...

There are four types of users you can create in Keap:

- **Admin**: This profile has full permission to access *every* aspect of your CRM system. Use this sparingly and only with trusted people

- **Limited Admin**: This role can add users (except **Admin** profiles), and they cannot change account setups

- **Manager**: This role can view everything except merchant accounts and *cannot* add users

- **Staff**: This role can only view their own contacts, tasks, and deals

By following this method of creating job roles within your CRM system, you can efficiently manage users, save time, and ensure continuity in your business operations. Instead of creating new profiles for each employee, you create defined roles and reduce the need for account deactivation, profile creation, and task reassignment when staff changes occur. This approach can ultimately save your business time and money in the long run.

Setting your business profile

Including your contact information at the bottom of emails is crucial for **CAN-SPAM compliance** because it allows recipients to easily identify and contact the sender, ensuring transparency and legitimacy in email communications. This legal requirement helps protect recipients from unsolicited or deceptive emails and fosters trust between businesses and their email recipients, contributing to a more secure and reputable email marketing ecosystem.

In this recipe, we'll cover how to optimize your business profile so that you not only comply with the necessary rules but also create consistent branding.

How to do it...

1. Start by clicking on your initials in the lower-left corner of the navigation.

2. Next, click on **Settings** in the pop-out menu. This will bring you to the **Settings** page, where you will choose **Business Profile**.

3. Completely fill in all your applicable data in the boxes provided:

    A. Your business name

    B. Choose your business type from the drop-down menu

    C. Your business email and phone number

    D. Your business website address (if applicable)

    E. Your country and complete address

4. Verify that the time zone is set to your preferred zone or click **Use my location** to detect your zone automatically:

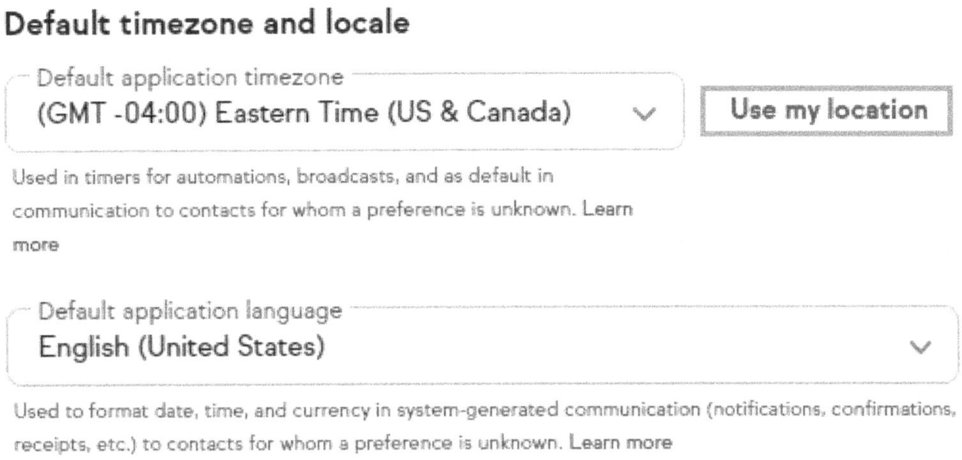

Default timezone and locale

Default application timezone
(GMT -04:00) Eastern Time (US & Canada)   ⌄   Use my location

Used in timers for automations, broadcasts, and as default in communication to contacts for whom a preference is unknown. Learn more

Default application language
English (United States)   ⌄

Used to format date, time, and currency in system-generated communication (notifications, confirmations, receipts, etc.) to contacts for whom a preference is unknown. Learn more

Figure 2.7 – Verifying your time zone and language

Drag your logo to the box or click to search your PC. When finished, click **Update business profile** to save your changes:

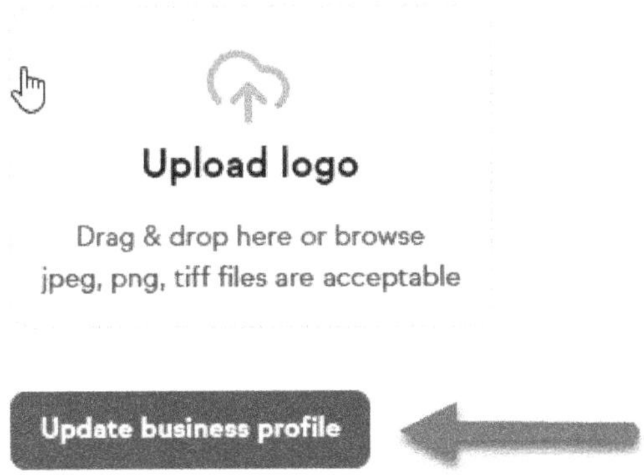

Upload logo

Drag & drop here or browse
jpeg, png, tiff files are acceptable

Update business profile

Figure 2.8 – Uploading your logo

Your logo is utilized in the creation of quotes, invoices, emails, landing pages, and various other materials. For optimal results, your logo should be a minimum of 300 px by 300 px in size.

How it works...

By following these steps, you will not only comply with CAN-SPAM regulations but also establish a professional and transparent identity for your business in email communications. Your recipients will easily identify and contact you, fostering trust and transparency in your marketing efforts. Additionally, the inclusion of your logo ensures consistency in your branding across various materials, contributing to a cohesive and recognizable brand identity.

Connecting to your calendar

Automating your appointment scheduling is one of the most powerful tools in your CRM system. When you allow people to self-serve, you take a lot of manual work off your plate. In this recipe, we are going to cover connecting your Keap calendar to your personal calendar, such as Google or Outlook. This is an important step to ensure a great experience for you and your potential lead or client.

How to do it...

The first step is to integrate your CRM system with your personal calendar so that you can avoid any overbooking or **double booking**. Overbooking or double booking occurs when you are using multiple calendars that are not synced, allowing one or more appointments to be scheduled for the same time:

1. We're going to start by clicking on your initials in the lower-left corner of the navigation.
2. Next, click on **Settings** in the pop-out menu. This will bring you to the **Settings** page.
3. Now select **Integrations**.
4. For this example, I will be choosing the **Google Calendar** option by selecting the **Connect** button:

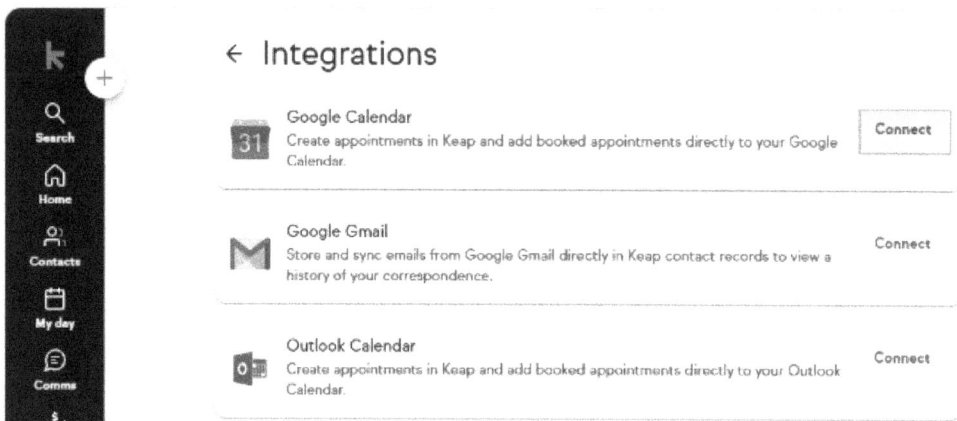

Figure 2.9 – Selecting your calendar option

5. Choose the profile of the calendar that you want to connect with:

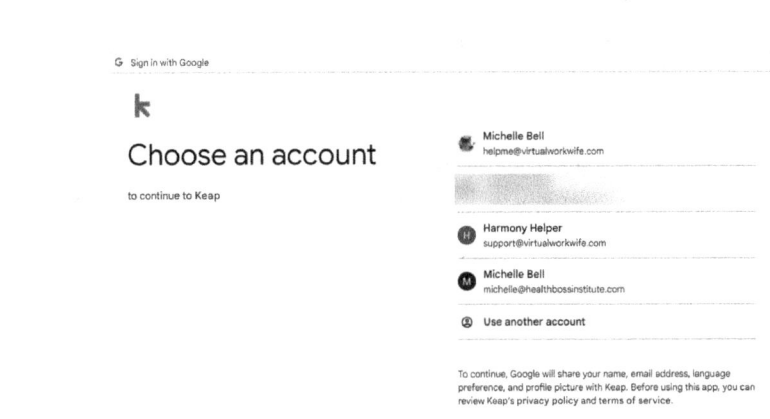

Figure 2.10 – Selecting your profile

6. Review the permission values being established and then click the **Allow** button.

7. Confirm your connection was successful by looking for the **Connected** notification in the **Google Calendar** box:

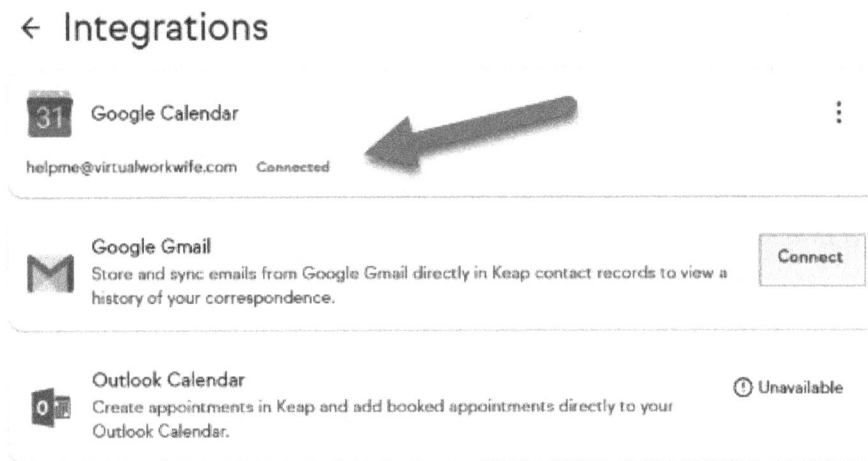

Figure 2.11 – Verify connection

How it works...

By following these steps, you lay the groundwork for creating scheduling links that will streamline your appointment scheduling process. Later, we will cover creating booking links, which allow your potential clients to automatically choose appointment times and populate your integrated calendar with those confirmed bookings directly.

This not only reduces your manual workload but also helps prevent overbookings and double bookings, ensuring that your time is used efficiently and that appointments are well organized. This feature can significantly enhance your CRM system's capabilities and lead to a more productive and organized approach to converting leads into paying clients and raving fans.

Connecting to email

In today's fast-paced world, effective communication with customers is paramount. Whether you're managing sales leads, providing support, or nurturing client relationships, integrating your email with Keap will revolutionize how you interact with your contacts. By seamlessly syncing your email account with your CRM system, you unlock a wealth of benefits that streamline communication processes, enhance your ability to deliver personalized experiences to your customers, and ensure that all customer interactions are automatically recorded, providing a comprehensive view of the customer journey.

In this recipe, we will take you through the connection and setup process step by step. Whether you're a seasoned user looking to enhance your system's capabilities or a newcomer eager to harness the power of integrated communication tools, this chapter provides the essential knowledge and practical guidance you need to seamlessly connect your email to your CRM system.

How to do it...

1. Start by clicking on your initials in the lower-left corner of the navigation.
2. Next, click on **Settings** in the pop-out menu.
3. Select **Integrations**.

4. For this example, I will be selecting **Google Gmail**. Click the **Connect** button:

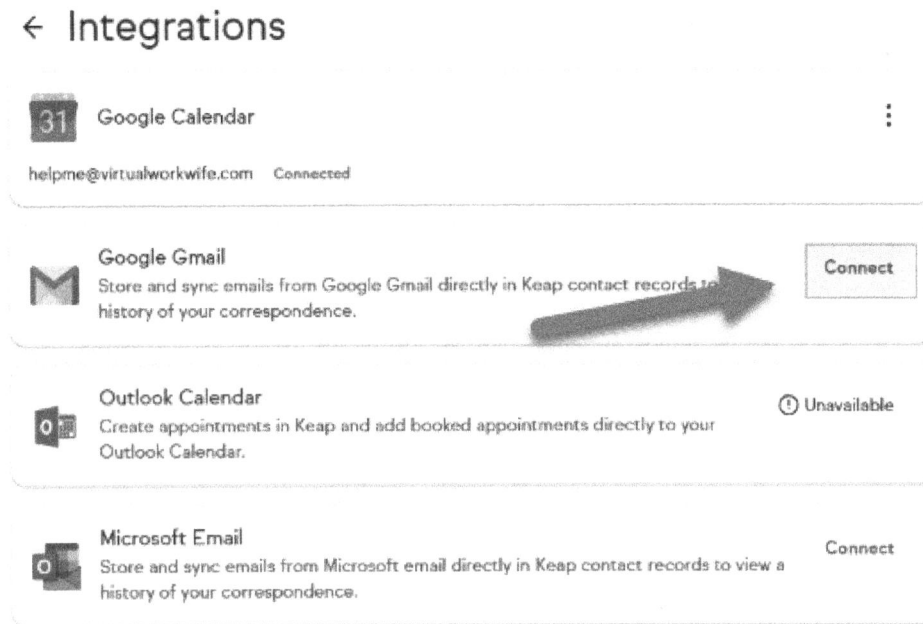

Figure 2.12 – Verify connection

5. Review the permissions being established and then click the **Allow** button.

6. Confirm the connection was successful by looking for **Refresh** in the **Google Gmail** box.

How it works...

By following these steps, you will establish a seamless connection between your email account and your CRM system, ensuring that all customer interactions are automatically recorded. This integration not only simplifies communication but also provides you with a comprehensive overview of the customer journey. It's a powerful tool for managing customer relationships, enhancing communication efficiency, and maintaining an organized record of all interactions with your customers.

Setting up appointment types

Now that your integrations are complete, we can begin to reap the benefits of having automated appointment scheduling! This is by far one of the easiest and most effective automations you can implement in your business.

How to do it...

1. Begin by selecting **MY DAY** in the navigation bar. By default, it will open the **Appointments** tab.

2. To create a new appointment, click the plus (+) sign:

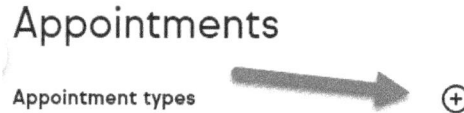

Figure 2.13 – Click the plus sign to create a new appointment

3. Next, you will give your appointment type a name.

4. Now let's define how you will meet with your contact:

   A. **Online**: Use this option for video calls, podcast recordings, and so on

      • Zoom (if integrated) creates a unique Zoom link for your call

      • Use an online meeting link (manually enter a link)

      • Ask attendees to use their online meeting link

   B. **Phone**: Use this when you want to have a personal call

      • Call the lead or client (ask the invitee for a phone number)

      • Ask the invitee to call (manually enter your number or check the box to use your Keap business line)

   C. **In-person**: Use this when you meet at a designated location

      • Choose a location now (manually enter an address)

      • Let the invitee choose the location (ask them for the address)

   For this example, I will be selecting **Online | My Zoom meeting link**:

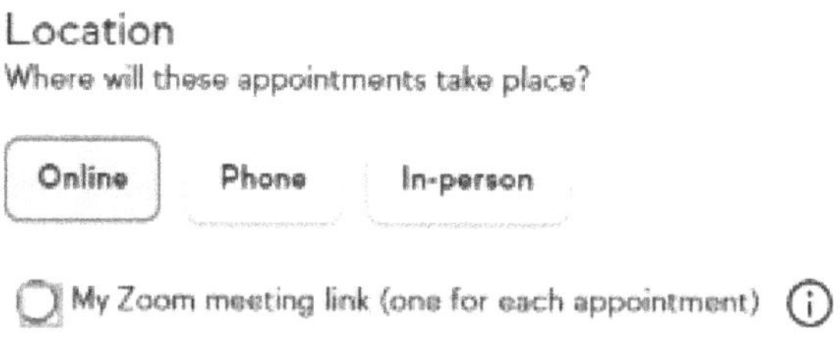

Figure 2.14 – Select appointment type

5. Set the duration of your call by selecting a timeframe from the drop-down menu. Use the **Buffer time** option if you want to give yourself time between calls:

Important note

If you have a busy schedule or other obligations over the course of a regular day, it is highly recommended that you don't allow anyone to create a same-day appointment. You can prevent this by setting **Advance Notice** to 24 hours.

Figure 2.15 – Set availability

6. Use the toggle button to open the editor and add any instructions or additional information to your appointment:

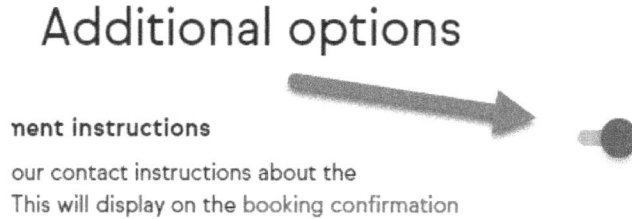

Additional options

nent instructions

our contact instructions about the
This will display on the booking confirmation

Figure 2.16 – Adding instructions to your appointment

7. Check the box next to all the calendars for which you want this appointment to determine your availability, and click the **Next** button to save your progress.

8. Click **Finish** to save your appointment.

How it works...

Your new appointment type will now be displayed in the left-side menu on **My Day**. Automatic appointment reminders have already been added for you. However, it is important to note that automatic appointment reminders represent a default value that is applied to all appointment types in Keap.

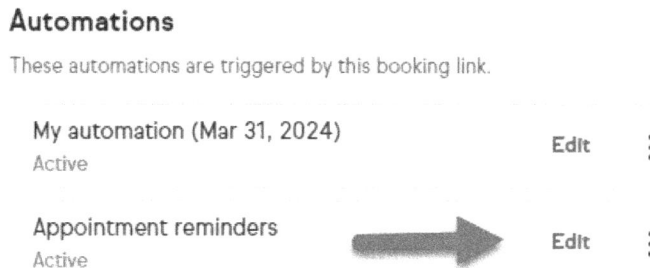

Automations

These automations are triggered by this booking link.

My automation (Mar 31, 2024)
Active   Edit   ⋮

Appointment reminders
Active   Edit   ⋮

Figure 2.17 – Editing appointment reminders

> **Important note**
> You can edit the reminders by clicking on the **Edit** button, but beware—you will be editing them for **ALL** appointments, not just the one you are currently working on.

Your availability should be designated based on the appointment type; for instance, you might choose to schedule only one discovery call each day, as these calls are typically offered free of charge.

By allocating specific time windows for different types of appointments on each day, you can prevent your calendar from becoming overloaded with one type of call over another, ensuring a balanced distribution of your time and resources.

Adding additional automation to your appointments can take a lot of manual tasks off your plate:

- Automate sending text reminders
- Create a deal in your pipeline so you never lose track of potential sales
- Apply tags to your appointment for segmentation and data tracking

All automations are based on the **When, Then, Stop** method. You will define when you want an action to take place, what that action will be, and what triggers the action to stop.

For example, creating a **DEAL** when an appointment is made is a visual way to keep track of all your activities. Think about the likes of Trello but with more automation! We'll look at deals and automation more in later chapters.

Payment processing

Before you begin collecting payments via Keap, you will need to connect to a merchant account. A **merchant account** is a specialized financial account that allows businesses to accept credit and debit card payments securely and efficiently.

To collect payments online, your merchant account acts as a bridge between you, your customer, and your payment processor. It facilitates the authorization, processing, and settlement of online transactions, ensuring that funds are transferred from the customer's account to the business's account securely and seamlessly.

Connecting a merchant account to your CRM system is a crucial step in establishing your online presence and facilitating transactions with your customers. While the process may seem daunting, understanding the key factors involved in choosing the right merchant account provider and setting up your account can simplify the journey and ensure smooth payment processing for your business.

This section does not provide a step-by-step guide to setting up a merchant account. It does, however, offer valuable insights and considerations to help you navigate the process with confidence. By understanding the factors involved in choosing a provider and setting up your account, you can make informed decisions that support the growth and success of your business in the digital marketplace.

Whether you already have a merchant account or need to set one up for the first time, Keap has options for you to explore. You can click on the **Learn more** options, as shown in the following screenshot:

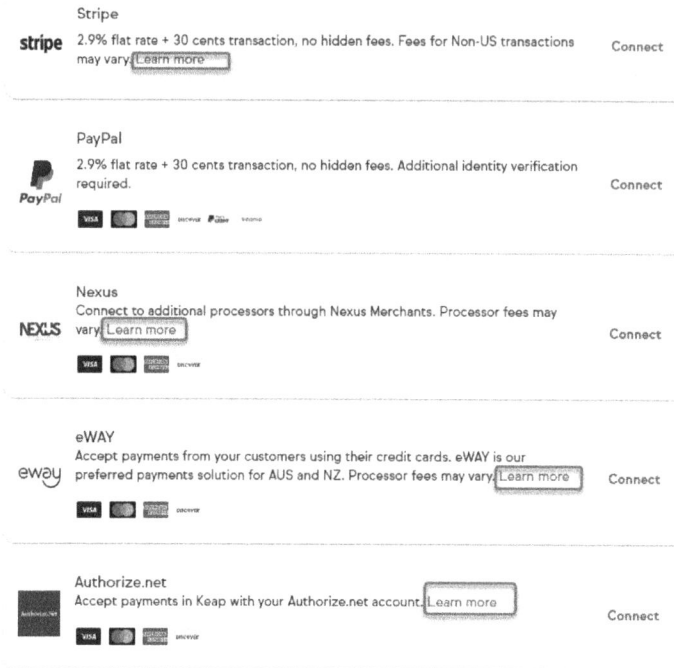

Figure 2.18 – Learn more options

It is essential to first assess your business needs and objectives before choosing a merchant. Consider factors such as the following:

- Transaction volume
- Types of payments accepted (e.g., credit cards and digital wallets)
- Desired features (e.g., recurring billing and invoicing)
- Applicable fees
- Budgetary constraints

By defining your requirements upfront, you can narrow down your options and focus on choosing a merchant account provider that aligns with your specific needs.

Keap has conveniently provided links to each merchant to help facilitate your research. Look for providers with transparent pricing structures, reliable customer support, and robust security measures to safeguard sensitive payment information.

Once you've selected a merchant account provider that meets your criteria, follow their onboarding process to set up your account. This typically involves submitting required documentation, such as the following:

- Business registration details

- Banking information

- Configuring account settings

While the specifics of account setup may vary depending on the provider, ensure that you adhere to any of the provided guidelines or requirements to expedite the process and avoid delays in activating your account.

Click the **Connect** button next to your preferred option and follow the onboard instructions.

Connecting your merchant account and automating payment collection is vital for business growth. It ensures efficiency, accuracy, and financial stability while saving time and enhancing the customer experience. Additionally, it provides valuable data insights, enhances security, and gives your business a competitive edge in the market. Embracing payment automation is a strategic move that paves the way for long-term success and expansion.

Creating products

In today's digital landscape, where clicks and swipes reign supreme, your CRM system is like the launchpad for a seamless online ordering experience that syncs perfectly with your customers' needs.

Quotes, invoices, subscriptions, and order forms are the building blocks of your online presence, creating a space where customers can effortlessly browse, engage, and snag what they're after. It's all about crafting an experience that's intuitive and inviting, and it all begins with your products!

However, we all know that the devil is in the details, right? That's why every little tweak, from organizing your product lineup to fine-tuning prices and keeping your inventory in check, and every little adjustment matters when ensuring a streamlined ordering experience.

When your products are neatly organized in your CRM system, it's like having a secret weapon for your marketing game. You can quickly launch low-ticket offers or create product bundles, analyze trends, and adjust your strategies on the fly to boost those sales figures.

Whether you're just dipping your toes into the online world or you're a seasoned pro looking to optimize your e-commerce, it's time to roll up your sleeves and unlock all that untapped potential waiting for you!

How to do it...

1. Begin by selecting **Sales** from the navigation menu and then clicking on **Products**:

    A. If this is your first time creating a product, click on the **Create a product or service** link in the center of the page.

    B. For all future product creation, click the **Add a product** button in the top-right corner of your screen:

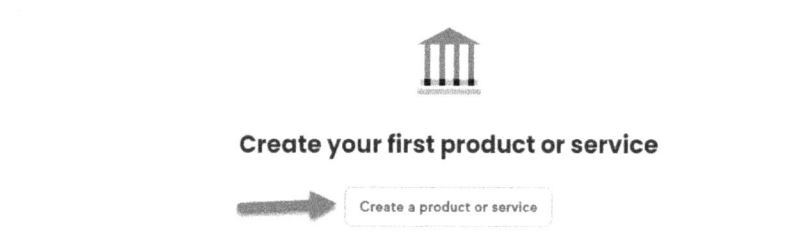

Create your first product or service

Create a product or service

Figure 2.19 – Adding products

2. Enter a name for your product.

3. Add a description. The description will be included in the receipt, so include as much information as possible to ensure your buyers can easily recognize your product.

4. Set a price for your product. Check the tax box and configure the tax rate(s) (if applicable) using the drop-down menu:

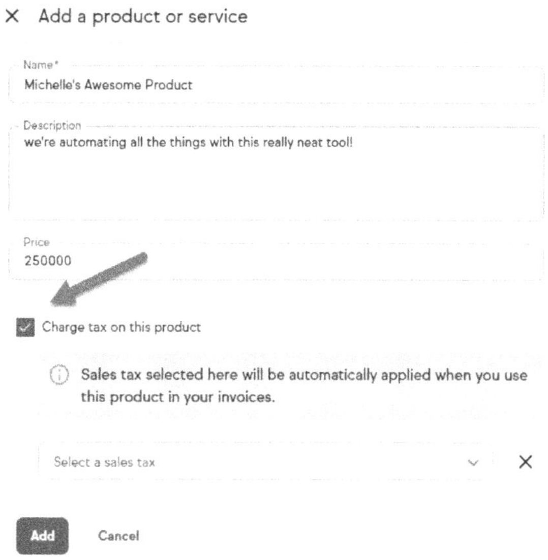

✕   Add a product or service

Name*
Michelle's Awesome Product

Description
we're automating all the things with this really neat tool!

Price
250000

☑ Charge tax on this product

ⓘ Sales tax selected here will be automatically applied when you use this product in your invoices.

Select a sales tax   ⌄   ✕

Add   Cancel

Figure 2.20 – Checking the tax box

5. To save your product to your inventory, click the **Save** button.

How it works...

Crafting products within your CRM system isn't just about checking boxes; it's a strategic move that simplifies ordering, streamlines fulfillment, amps up efficiency, and ensures a hassle-free journey for your customers. Plus, it lays the groundwork for creating automation that can take your sales from single orders to repeat purchases.

By following the previous steps, you're setting the stage for a thriving e-commerce presence, personalized marketing, data-driven decisions, and an overall smoother customer experience.

Zapier integration

While Keap offers many onboard integrations, there are still some tools you may have to use that use third-party software to connect to Keap.

Zapier is an online automation tool that allows users to connect and integrate various web applications and services, enabling them to automate repetitive tasks and workflows without any coding or technical expertise. Users create **Zaps**, which are automated workflows that connect two or more apps, triggering actions in one app based on events or actions in another.

How to do it...

For example, you can create a Zap that automatically posts a notification in Slack when someone purchases a product in Keap:

1. Log in to Zapier.
2. Navigate to **My Apps** from the navigation bar.
3. Click on **Add Connection** in the top-right corner of your screen.
4. In the search box, type Keap.
5. In the pop-up window, use your login credentials to connect to your Keap account.

How it works...

With Keap-Zapier integration successfully set up, you can now begin creating automations that connect your Keap CRM system with various external tools and services. This integration offers tremendous flexibility and efficiency in managing your business processes. Keap also provides premade Zaps to assist you in the automation process, making it easier to set up and customize workflows tailored to your specific needs.

There's more...

The integration of Keap and Zapier is a valuable asset for businesses looking to streamline their operations, increase efficiency, and expand their CRM system's functionality:

- **Seamless workflow automation**: Create Zaps to automate tasks and workflows between systems to reduce manual effort and improve operational efficiency

- **Integration with external tools**: Connect Keap to a wide range of third-party applications, expanding your CRM system's capabilities and versatility

- **No coding required**: You can set up complex automations without the need for coding or technical expertise, making it accessible to users of varying skill levels

- **Customizable workflows**: Tailor Zaps to match your business requirements and automate processes that are specific to your industry or operations

- **Increased productivity**: Automation frees up time and resources that can be allocated to more strategic tasks, helping you achieve higher productivity

- **Enhanced data flow**: Ensure data consistency and accuracy by automating data transfers between your CRM system and other apps

- **Streamlined communication**: Create Zaps that facilitate communication between different platforms, such as notifying your team in Slack when specific events occur in Keap

With this integration, you can create automated workflows that connect Keap to a wide range of third-party applications, all without the need for technical expertise. This not only simplifies your day-to-day tasks but also provides a significant competitive advantage by allowing you to focus on strategic initiatives and enhance your business's overall performance.

Google reviews

By adding Google reviews to your dashboard widgets, you can automatically capture and store customer reviews and ratings, allowing your team to track and respond to feedback more effectively. This integration streamlines the process of monitoring customer satisfaction, helping you improve your products or services and enhancing overall customer relationships.

How to do it...

1. From your dashboard, scroll down until you see the **Reviews** widget:

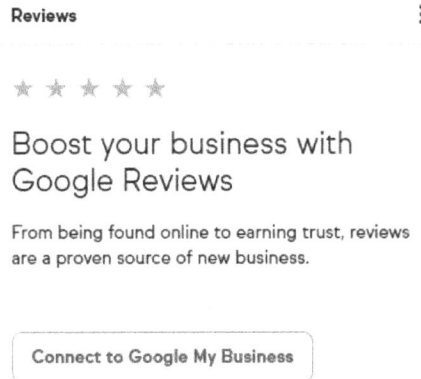

Reviews   ⋮

★ ★ ★ ★ ★

Boost your business with Google Reviews

From being found online to earning trust, reviews are a proven source of new business.

Connect to Google My Business

Figure 2.21 – Google Reviews widget

2. Click on the **Connect to Google My Business** link on the widget.

3. Choose the profile you want to connect to.

4. Review the permission values being established and then click the **Allow** button.

How it works...

You can now copy the review link and use it in your automations or directly request a review by clicking on the **Request a review** button and selecting a contact.

Adding the link to your automations is the easiest way to consistently get reviews without feeling the pressure that comes from actually asking for them!

3
Managing Contacts

In the world of business, the effective management of leads and clients is paramount to success. Your **customer relationship management** (**CRM**) system is the cornerstone of your business management. This chapter will guide you through the essential steps of contact management within Keap, empowering you to organize, segment, and leverage your contact data for enhanced business outcomes.

The process of managing contacts in your CRM system begins with the foundational task of adding contacts manually. Whether it's inputting new leads acquired through networking events or manually entering client details, this step ensures that your CRM system contains accurate and comprehensive contact information.

Keap uses a system of tagging contacts to provide a dynamic way to categorize and segment your list based on specific criteria such as industry, purchase history, or location. This enables targeted communication and personalized engagement strategies tailored to the unique needs of different segments within your audience.

In this recipe, we will cover importing contacts from your external systems to ensure that no valuable leads or client information is left behind. Whether you're migrating from another CRM platform or consolidating contacts from various sources, this step streamlines data management and centralizes your contact database within your CRM system.

We will also be adding custom fields to your contact records, enabling you to capture and track additional information specific to your business needs. Whether it's recording preferred communication channels, tracking referral sources, or capturing unique identifiers relevant to your industry, custom fields enhance the depth and granularity of your contact data.

By mastering these essential aspects of contact management within your CRM system, you'll unlock the full potential of your customer data and harness its power to drive business growth and foster stronger customer relationships. From efficient data organization to targeted communication strategies, effective contact management lays the foundation for success in today's competitive business landscape.

In this chapter, we will cover the following recipes:

- Adding contacts manually

- Adding companies

- Grouping contacts

- Importing contacts

- Adding custom fields

- Introduction to tagging

- Searching contacts

Technical requirements

For this chapter, the following is required:

- A contact list in the .csv format (for uploading). If you do not have one, you can create a sample list to learn how importing works.

Getting ready

Later on in this chapter, you will learn how to import contacts into your CRM system. If you haven't done so already, you will need to export your contact list(s) from your current marketing tools or contact managers. Be sure to save them as .csv files so they can be uploaded to Keap.

> **Important note**
> For more information on exporting and importing contacts, watch this video: www.virtualworkwife.com/contacts-import-export.

Adding contacts manually

Adding contacts to Keap is a relatively simple process, making it convenient to include potential leads or clients manually. Keap's user interface is designed for efficiency, allowing for the swift input of essential information, such as names, contacts, and relevant interactions. Recording all interactions is a practical approach to maintaining an accurate history of your client acquisition efforts, which requires all contacts to be in the CRM system. This method ensures a systematic record of engagements, aiding in organized and automated follow-up strategies for optimal efficiency.

In this recipe, we will walk through the steps to create contacts and companies in your Keap CRM system.

How to do it...

1. Click on the **CONTACTS** tab in the left-side navigation bar to open the menu and choose **People**.

2. Keap will display a list of contacts in your CRM system and ask you to add a contact or import contacts:

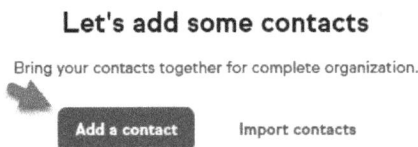

Figure 3.1 – Adding contact options

3. Click on the **Add a contact** button to open the pop-up box.

4. Click the **Lead**, **Client**, or **Other** indicator at the top of the box to indicate what type of contact you will be adding. This is your first step in segmenting your list!

5. In the pop-up box, enter the contact information, filling in as much detail as you can. Remember, the more information you can provide, the better you will be able to segment your list later on:

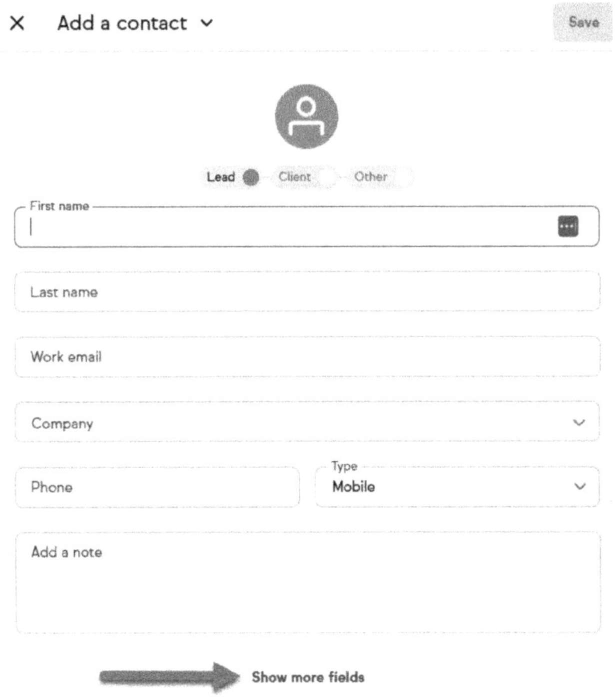

Figure 3.2 – Adding contact data

6. Clicking the **Show more fields** link will display additional fields for contact information, such as alternative phone numbers, shipping/billing addresses, and custom fields.

7. Click **Save** in the upper-right corner of the pop-up box to save your new contact.

How it works...

When you've added a contact to Keap, you will immediately be offered options for how to reach out to them. While these email templates are prewritten, you can edit the content before you send the email:

Take a peek at your new lead.

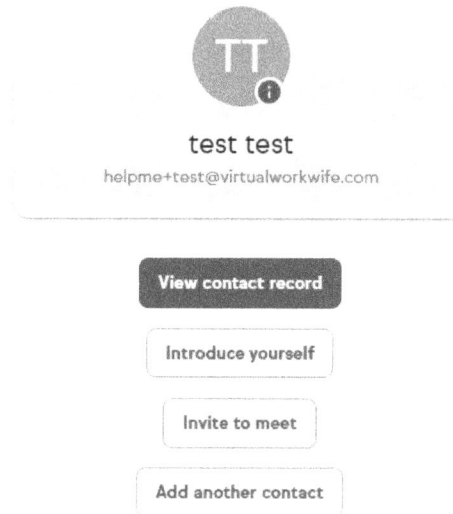

test test

helpme+test@virtualworkwife.com

View contact record

Introduce yourself

Invite to meet

Add another contact

Figure 3.3 – Options for working with contacts

-   **View contact record**: This closes the window and displays the full contact card onscreen

-   **Introduce yourself**: This opens the email editor and a prewritten welcome email

-   **Invite to meet**: This opens the email editor and a prewritten invitation to meet

-   **Add another contact**: This opens the **Add a contact** popup again

Taking the time to reach out to a new lead holds immense value in terms of building meaningful connections and fostering potential business relationships. By reaching out, you demonstrate genuine interest, can address specific needs, and create a positive first impression. Your effort not only establishes a foundation of trust but also opens the door for effective communication. Investing time in reaching out to a new lead is an invaluable step toward understanding their unique requirements, tailoring your offerings, and ultimately increasing the likelihood of converting a lead into a loyal client.

Adding companies

Managing relationships sometimes extends beyond individual contacts to encompass the companies they belong to. Adding company records to your CRM system offers yet another avenue for enhancing business growth. By centralizing company data, you gain a comprehensive overview of your client base, allowing for more informed decision-making. Tracking your interactions and engagements with companies becomes more efficient, providing insights into their preferences and needs. This, in turn, enables targeted communication and personalized services.

A detailed company profile within your CRM system fosters collaboration among team members, ensuring everyone is on the same page regarding the status of various accounts. This recipe will explore the crucial steps involved in adding companies to your CRM and how to connect people (contacts) and companies.

How to do it...

1. Click on the **CONTACTS** tab in the left-side navigation bar to open the menu and choose **Companies**.

2. Keap will display a list of companies in your CRM and ask you to add a company:

Keep your contacts organized by company

See who you work with at each company, and manage it all in one place.

Add a company

Figure 3.4 – Adding a company options

3. Click on the **Add a company** button to open the pop-up box.

4. In the pop-up box, enter the company information, filling in as much detail as you can. Remember, the more information you can provide, the better you will be able to segment your list later on.

5. Clicking the **Show more fields** link will display additional fields for storing contact info, such as alternative phone numbers, shipping/billing addresses, and custom fields.

6. Click **Add company** in the lower-left corner of the pop-up box to save your new company.

7. Click the **View company record** button to close the pop-up box:

Take a peek at your new company

virtualworkwife

View company record

Add another company

Figure 3.5 – View a company record

At this point, you can connect contacts to your company record. Linking contacts to companies in a CRM system enhances business insights by establishing contextual relationships. This connection allows for a holistic view of client interactions, streamlines communication, and facilitates targeted strategies, ultimately fostering stronger and more fruitful relationships.

8. On the company record, begin by clicking the **Add a contact** button to open the contact search box.

9. From the **Select contacts** drop-down menu, you can either scroll down your list of contacts and click to select them or type in the box to filter the list:

✕ Add contacts to this company
virtualworkwife

Select contacts

test bell ⊗   test test ⊗   Tech Hero ⊗   ⌄

Add contacts   Cancel

Figure 3.6 – Adding contacts to a company

10. Click the **Add contacts** button to connect the selected contacts to the company.

How it works...

Think of contacts and companies in a CRM system as being like best friends who always share everything. When they sync up, it's like having a smooth flow of information; every person (contact) is linked to a specific group (company). In this way, you get the full picture of who's who in the organizational landscape. Syncing contacts and companies makes communication super easy, helps teams work together seamlessly, and lets you handle customer relationships like a pro. It's basically the secret sauce that makes managing relationships a whole lot simpler and more effective!

Grouping contacts

Grouping contacts in a CRM system is like creating organized squads for your network. Instead of dealing with each person separately, you can put them into specific groups based on common traits, interests, or interactions. In this way, when you need to reach out, you can target a whole group at once, making communication more efficient. Whether it's a marketing campaign or checking in with a particular segment, grouping contacts simplifies your CRM game, helping you stay on top of your connections and tailor your approach for maximum impact.

In this recipe, we will begin segmenting your list by adding criteria to create groups.

Getting ready

Before you begin creating groups, you first need to understand how grouping works.

Groups are based on criteria. The process involves bringing together similar elements to create a distinct category or set, making it easier to manage and analyze information in a more structured and meaningful way. In various contexts, grouping can be applied to data, contacts, tasks, or any set of items, allowing for more efficient organization and streamlined operations.

Within the **Contacts/Company** area of Keap, you can group people using the following criteria:

- **Equals**: Items in this group share an identical value for the specified criterion. For example, contacts where the job title is CEO.

- **Not equals**: Items in this group have different values to that of the specified criterion. For instance, contacts where the first name isn't Harry.

- **Contains**: This criterion includes items that contain the specified term within their content. For instance, companies with names that contain "consulting."

- **Does not contain**: Items in this group lack the specified term within their content. For instance, contacts whose email addresses do not contain "@companyname.com."

- **Starts with**: Items in this group begin with the specified term. For example, filtering contacts whose names start with A.

- **Ends with**: This criterion includes items that conclude with the specified term. For instance, companies with names that end with "Inc."

- **Is empty**: This filter includes items that do not have any content or information for the specified criterion. For example, contacts without a specified phone number.

- **Is filled**: This criterion encompasses items that have content or information for the specified criterion. For instance, filtering tasks that have a filled-in due date.

- **Includes any**: This criterion gathers items that have at least one of the specified values or conditions. For example, contacts whose tags include any of the terms prospect, client, or lead.

- **Includes all**: This filter narrows down items to those that meet all specified values or conditions. For instance, tasks that include all of these tags: "Urgent," "Priority," and "Follow-up."

- **Excludes any**: This criterion filters out items that have at least one of the specified values or conditions. For example, filtering contacts whose categories exclude either of the terms "Inactive" or "Closed."

- **Excludes all**: This filter excludes items that meet all specified values or conditions. For instance, excluding deals that have all of these tags: "Completed," "Closed," and "Canceled."

How to do it...

1. Click on the **CONTACTS** tab in the left-side navigation bar to open the menu and choose **Groups**.

2. Keap displays a list of groups in your CRM system. Click the **Create a group** button to open the pop-up box:

Categorize your contacts with groups

Take the right action fast. Send a tailored broadcast, add them to an
automation, add or remove tags, or even export a group.

Create a group

Figure 3.7 – Adding groups

3. Next, click in the **Add a filter** box and select a criterion you want to use to filter your list. For this example, I will use **Job title** | **Equals** | CEO.

4. Continue adding filters until you have the desired criteria:

> **Important note**
> The more filters you add, the more narrowed down your list becomes. You can include or exclude people from your group by adding more or fewer filters.

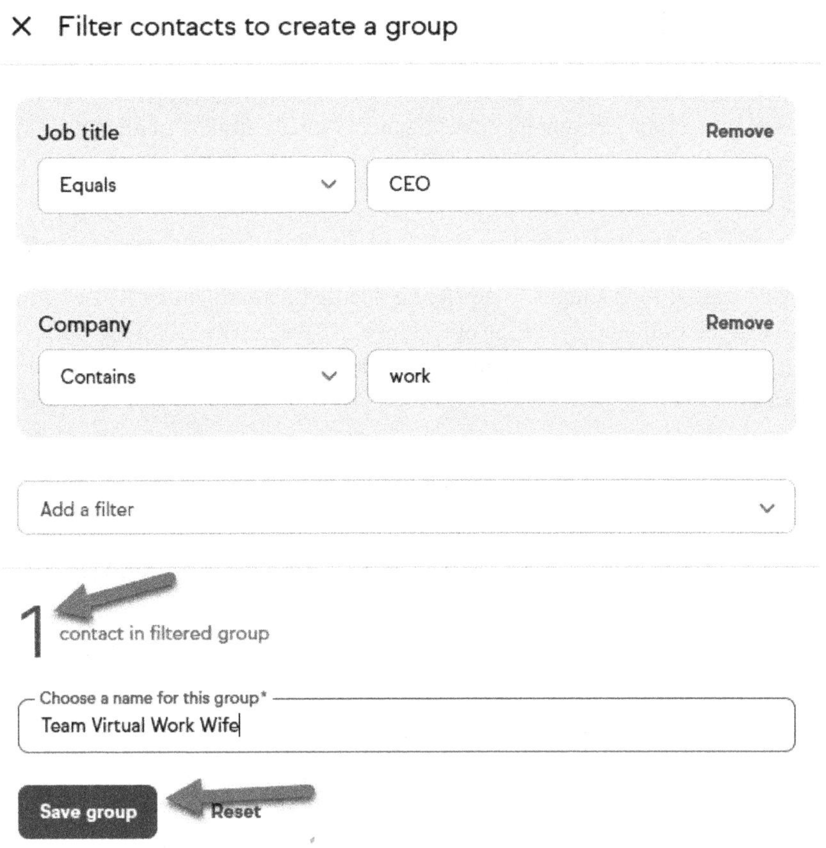

✕ Filter contacts to create a group

Job title			Remove
Equals	⌄	CEO	

Company			Remove
Contains	⌄	work	

Add a filter	⌄

1 contact in filtered group

Choose a name for this group*
Team Virtual Work Wife|

Save group   Reset

Figure 3.8 – Setting your group criteria

5. Add a name for your group in the provided box.

6. Click the **Save group** button.

How it works...

Keeping a clean and systemized CRM list by using group criteria offers substantial advantages. This organizational approach streamlines daily tasks, reduces search times, and enhances the efficiency of your operations.

Grouping or segmenting contacts in your CRM system allows for targeted communication through broadcasts or incorporating groups into your automations. This approach enables more personalized outreach, resulting in more effective marketing and a tailored customer experience.

Additionally, the ease of collaboration among team members, coupled with improved analysis and reporting capabilities, ensures that your business is well equipped to adapt, scale, and make informed decisions based on accurate, real-time data. In essence, a well-maintained CRM system with clear group criteria is a foundational element for achieving operational excellence and delivering superior customer satisfaction.

Importing contacts

Ensuring the cleanliness of your contact lists before uploading them to your CRM is a fundamental step in maintaining the health and efficiency of your system. Beyond avoiding the clutter caused by duplicate entries and inaccuracies, a clean contact list contributes to accuracy and reliability when it comes to reporting data.

Having accurate data is pivotal for making informed business decisions and fostering trust in your customer relationships. By eliminating errors and ensuring up-to-date information, your team can confidently engage with contacts, leading to more effective communication and personalized interactions.

Removing outdated or inaccurate information not only reduces the risk of privacy breaches but demonstrates a commitment to ethical data management practices. In essence, the cleanliness of your contact lists sets the stage for a reliable, compliant, and efficient CRM system, positioning your business for success.

Once you have a clean and accurate `.csv` list, you can import those contacts into Keap.

How to do it...

1.   Click on the **CONTACTS** tab in the left-side navigation bar to open the menu and choose **People**.

2.   Keap will display a list of contacts in your CRM and ask you to add a contact or import contacts:

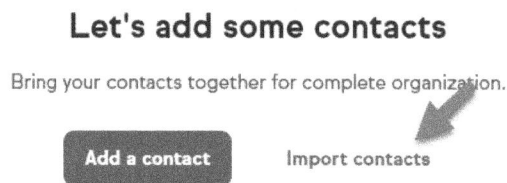

Let's add some contacts

Bring your contacts together for complete organization.

Add a contact   Import contacts

Figure 3.9 – Importing a list of options

3.   Click the **Import contacts** option.

4. There are many options for importing from other tools. For this exercise, we will choose the **My spreadsheet** option:

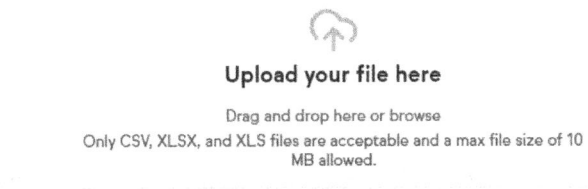

Upload your file here

Drag and drop here or browse
Only CSV, XLSX, and XLS files are acceptable and a max file size of 10 MB allowed.

Figure 3.10 – Uploading your list

5. Use **Drag and drop here or browse** to upload your file.

6. Keap will try to match your field (column) names to their standard fields.

    A. When a field is not matched, you will be given the chance to do one of the following:

    • Manually match it to a desired field

    • Add it as a custom field

    • Create a tag from the field value

    • Choose *not* to import that specific field data

7. When all your fields are properly mapped, click the **Next** button in the top-right corner to continue.

8. Indicate that you have permission to email your list by checking **Yes**:

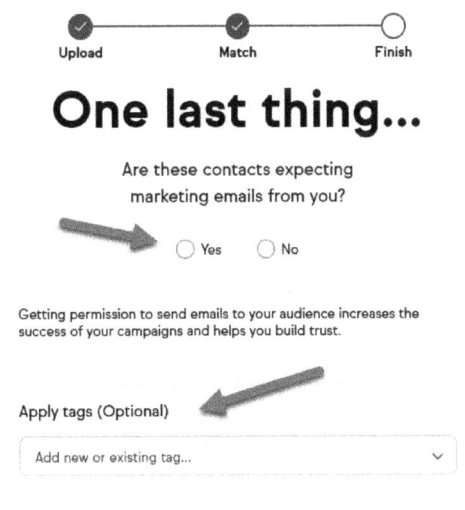

Upload   Match   Finish

One last thing...

Are these contacts expecting marketing emails from you?

○ Yes   ○ No

Getting permission to send emails to your audience increases the success of your campaigns and helps you build trust.

Apply tags (Optional)

Add new or existing tag...

Figure 3.11 – Tagging your new list

9. At this stage, you can begin to segment your list by giving it one or more tags:

    A. Click in the box and type to search or scroll your tag list.

    B. Choose an existing tag by typing or scrolling to locate one. You may also click on the + **ADD** link to create a new tag.

> **Important note**
> Applying a tag during the import process is a highly useful means of triggering automation.

10. Click the **Finish Import** button in the top-right corner to upload your list.

How it works...

By following the steps outlined in your recipe, you'll ensure your CRM is populated with clean, organized, and accurate contact data, paving the way for effective relationship management and strategic marketing efforts.

Adding custom fields

Custom fields allow you to tailor Keap to handle the specific information requirements of your business. Standard fields in a CRM system capture general data, such as names, email addresses, and phone numbers, but custom fields allow you to go beyond these basics.

Customization ensures that your CRM system becomes a personalized tool, reflecting the nuances of your business operations and enabling more precise tracking, analysis, and communication. Essentially, custom fields empower your CRM system to evolve with your business, accommodating its growth and adapting to the changing demands of your industry.

In this recipe, we will learn how to create custom fields. By incorporating custom fields, you can gather and organize industry-specific details, track unique customer preferences, or record data pertinent to your unique products or services.

Getting ready

Before you begin, it's important to understand how CRM systems handle data. The type of field you select should align with the specific data you intend to store, addressing your ultimate objectives for using those data, such as reporting or using it as a merge field to insert the stored value into an email.

It's important to note that once a field type is established, it cannot be changed to another type. For instance, if you initially decide to create a textbox and then later decide you need a drop-down box, you will need to create a new field and delete the old one.

There are four basic field types:

1. **Text and number fields**: These are best leveraged to store diverse custom information and are commonly applied as reference points and merge fields in emails, letters, and task templates:

   A. **Text**: This field accommodates all data types (letters, numbers, and symbols) but has a maximum capacity of 255 characters, making it ideal for brief responses.

   B. **Text Area**: This has a capacity of approximately 65,000 characters or around 9,000 English words; this field suits lengthy responses and is perfect for open-ended inquiries in web forms such as surveys or contact forms.

   C. **Whole Number**: This field exclusively stores whole numbers, rejecting letters, symbols, or decimal points.

   D. **Decimal Number**: This field stores decimal numbers and rounds up to the nearest hundredth value.

   E. **Percent**: While this field can store any numerical value, it visually displays numbers with a decimal point and a percentage sign (%).

   F. **Currency**: This field is reserved for numbers featuring a decimal point and is denoted by a dollar sign ($); this field is suited to monetary values.

2. **Date fields**: These are utilized to store personalized date details, such as birthdays, significant occasions, renewal dates, and so on. These dates can be integrated into emails, letters, and task templates to add a personalized touch and serve as event triggers:

   A. **Date**: This records a user-defined date. Input the preferred year manually or utilize the up and down arrows for selection.

   B. **Day of Week**: This saves the name of a day in text format, such as Tuesday.

   C. **Month**: This saves the name of a month in textual format, such as *May*.

   D. **Year**: This stores a four-digit year, such as *2024*.

3. **Option lists**: When creating criteria for searching and reporting, opt for option list fields. These fields entail defining a finite set of options for each category rather than permitting clients to input their own values:

   A. **Radio button**: This displays all available options at once; however, only one option can be selected at a time.

   B. **Drop-down**: This offers a list of options in a collapsible menu. Users must click to reveal the options from which to choose. Only one selection is allowed at a time. Each drop-down list can accommodate up to 2,000 characters or approximately 499 items.

   C. **Checkboxes**: This creates a list of options that can be checked. Unlike radio buttons, multiple checkbox options can be selected at once.

   D. **User**: This lists the names of your team members who have access to Keap.

4.   **Specially formatted fields**: Use these fields when you need to collect more data about your leads and clients:

    A.   **Email Address**: This stores additional email addresses and equips them with the send email icon for easy communication.

    B.   **Phone Number**: This stores additional phone numbers and makes them accessible to click-and-call devices and texting.

    C.   **Website**: This converts a web address into a hyperlink for one-click access.

    D.   **State**: This creates a drop-down list of states that you can populate to a list.

    E.   **Yes/No**: This is similar to a radio button but only has two options: yes and no.

> **Important note**
> It is possible to export field data and then import it back into the CRM system to move it from one field to another, facilitating a smooth transition between field types.

How to do it...

1.   Begin by clicking on your avatar in the lower-left corner of your CRM and then choosing **Settings**.

2.   Scroll down the options list and choose **Custom Fields**.

3.   Next, choose what you want to manage (e.g., the contact or company fields). For this recipe, we will be using **Custom contact fields**:

Figure 3.12 – Custom fields

4.   Click the **Create new** button in the top-right corner of your screen or the + icon to open the pop-up box.

5.   Name your custom field.

6.   Select your field type from the drop-down box.

7. Select whether you want the custom field to always be visible when you are looking at the contact record. You can manually update this setting later if you decide it is a critical data point.

8. Click the **Create field** button to save your new field.

Your custom field will now be visible and searchable on the left side of the field management page:

1. To add another custom field, you can now click on the + icon above the list of fields.

2. To edit a custom field name, do the following:

    I.   Click on the field to display its details.

    II.   Change the name as needed.

    III. Click the **Save changes** button to save.

3. To delete a custom field, do the following:

    I.   Click on the field to display its details.

    II.   Click the **Delete field** button.

    A warning message will appear, letting you know whether it is safe to remove the field or whether doing so will also cause you to lose the data stored in the contact records:

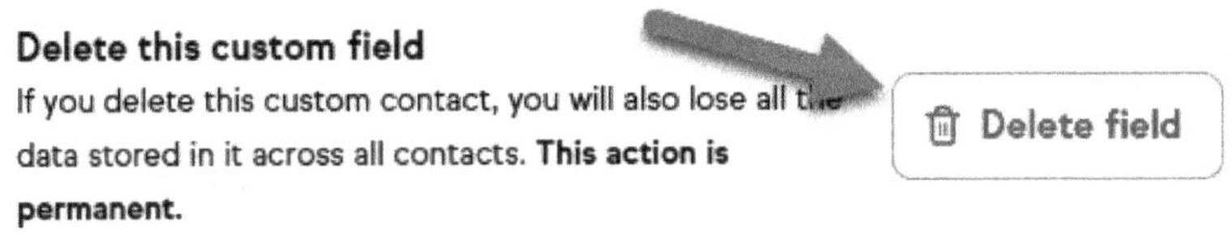

Figure 3.13 – Deleting a custom field

> **Important note**
> This action is permanent and cannot be undone without seeking help from Keap at significant expense!

4. Select **Delete field** to confirm your action.

How it works...

By taking these steps to create custom data fields, you can begin to build a fuller picture of your leads and clients. Custom fields help capture the specific details that matter to your business, giving you a better grasp of individual preferences and industry specifics. This knowledge empowers you to tailor your interactions, making your relationships stronger and boosting overall customer engagement.

Introduction to tagging

Tags in your CRM system are like digital labels that help you organize and categorize your contacts, companies, or deals. They act as quick identifiers, allowing you to easily group and filter information based on shared characteristics, interests, or stages. Using tags is essential because they make it a breeze to find specific groups of contacts, track important trends, and run targeted campaigns. It's like having a set of personalized shortcuts that streamline your CRM tasks, ensuring you can quickly access and manage your data in a way that suits your business needs.

Once you've established your tagging structure and begun applying tags to contacts, you will easily be able to perform quick, bulk actions with your tagged contacts, but Keap also provides a dedicated tag management page. We will also begin to send a broadcast to or export tagged contacts.

Getting ready

Before you begin, consider designing a well-defined naming and organization structure for your tags. This is crucial for several reasons:

- **Standardization**: Consistency simplifies data entry and retrieval, as users can easily identify and select appropriate tags without confusion. Consistent tags also facilitate accurate reporting and analysis by providing uniform data categories.

- **Easy navigation**: Clear and intuitive tag names enhance the usability of your CRM system, making it easier for users to locate relevant information. When tags are organized logically and named descriptively, users can quickly identify the purpose and relevance of each tag.

- **Facilitates search and filtering**: A well-organized tag structure facilitates the efficient searching and filtering of data. Users can use tags as search tools to retrieve records that match specific attributes or criteria. With a logical organization and naming convention, users can construct complex search queries using tags to precisely filter data and retrieve relevant information.

- **Enhanced reporting and analysis**: Tags play a vital role in categorizing and segmenting CRM data for reporting and analysis purposes. A well-structured tag system enables you to generate insightful reports and analyze trends based on tagged attributes.

- **Scalability and growth**: As businesses grow and evolve, having a scalable tag structure becomes increasingly important. A well-designed tag system can accommodate new categories, attributes, or changes in business processes without disrupting existing workflows. Scalability ensures that the CRM remains adaptable to evolving business needs and can continue to support efficient data management as the organization expands.

Overall, having a good naming and organization structure for your tags is essential for optimizing data management, improving user productivity, and enabling informed decision-making. By implementing a standardized approach to tagging names, businesses can leverage the full potential of their CRM system to drive success and growth.

How to do it...

In this recipe, we will learn how to navigate the tag management page, where you can quickly add or delete tags.

Working with tags

1. Begin by clicking on your avatar in the lower-left corner of your CRM system and then choose **Settings**.
2. Scroll down the options list and choose **Tags**.
3. You can add to or filter your tag list using the tools at the top of the page:

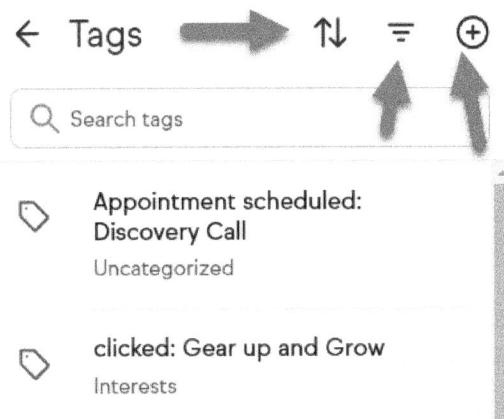

Figure 3.14 – Working with tags

4. Click on the tag you want to work on to display its attributes in the main window. From here, you can do the following:

   - Edit the tag
   - View and/or edit automations using this tag
   - View and/or edit contacts using this tag
   - Manually add a new contact and tag them with this tag
   - Select contacts with this tag and perform mass actions:

     - Send a broadcast email
     - Add the selected contacts to an automation
     - Add or remove them from another tag
     - Update their contact type

- Export them as a `.csv` file
- Delete the selected contacts

Editing a tag

For the purpose of this introduction to tagging, we will begin with editing a tag:

1. Select a tag from the list to display it in the main window.
2. Click the **Edit tag** button to open the pop-up box:

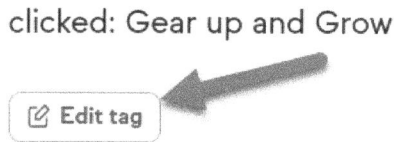

clicked: Gear up and Grow

☑ Edit tag

Figure 3.15 – Editing a tag

3. Edit the name.
4. Update the tag category via the following:

    I.   Select one from the drop-down menu.

    II.   Click the + **Add** button and create a new category.

5. Add a description of your tag.
6. Click the **Save** button.

Deleting a tag

To delete a tag, do the following:

1. Select a tag from the list to display it in the main window.
2. Click **Edit Tag**.
3. In the bottom-left section of the pop-up box, click the **Delete tag** button.

> **Important note**
> There is no warning message when you choose to delete a tag! This action is final and cannot be undone.

Applying a tag

To apply a tag, do the following:

1. Click on the **CONTACTS** tab in the left-side navigation bar to open the menu and choose **People**.

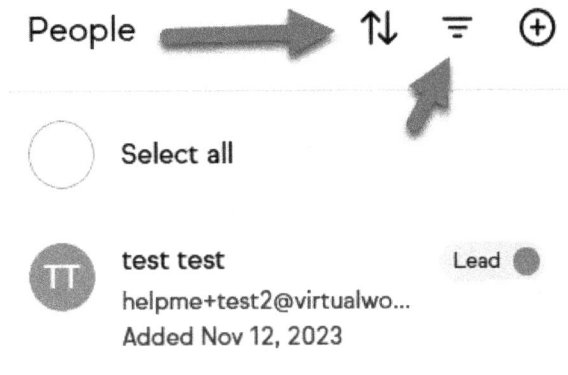

Figure 3.16 – Applying tags

2. Keap will display a list of contacts in your CRM system. Use the filter options to narrow your list down to those you want to tag:

   A. Check the **Select all** option to choose everyone, or click the icon next to the names of the people to select only those you want to tag.

3. Once a segment of contacts has been selected, a menu of buttons will appear at the bottom of the screen.

4. Click the **Tag** button to open the option box.

5. Type the tag name or scroll through the drop-down menu to select your tag(s).

6. Click the **Apply tag** button to confirm.

Removing a tag

To remove a tag, do the following:

1. Repeat *steps 1–6* from the *Applying a tag* section.

2. Click the **Remove** button to confirm.

How it works...

Tagging is one of the—if not the—most important roles in a CRM system. It involves categorizing records with descriptive labels to facilitate organization, search, and analysis. These tags help users quickly identify and retrieve relevant information, streamline data management processes, and enable targeted communication and reporting. A well-implemented tagging system enhances CRM usability, supports data-driven decision-making, and contributes to overall efficiency and effectiveness in customer relationship management.

Removing tags when appropriate keeps your CRM tidy. When you clear out unnecessary or outdated tags, you're decluttering your contact list. This simple step ensures that you only keep the tags that really matter, making it easier to find and understand your contacts. Monthly routine list maintenance is the best thing you can do for your CRM, ensuring it is accurate, organized, and ready for efficient use.

Searching contacts

Searching for contacts in your CRM using field values and tags is a straightforward and easy process.

Combining field values and tags allows for precise searches, helping you locate and manage contacts based on their attributes or engagement status. This functionality not only makes finding information quicker but also enhances your ability to run targeted campaigns or track specific groups within your CRM system.

How to do it...

1. Click on the **CONTACTS** tab in the left-side navigation bar to open the menu and choose **People**.
2. Keap will display a list of contacts in your CRM system. Use the filter options to narrow your list down to only the ones you want to see:

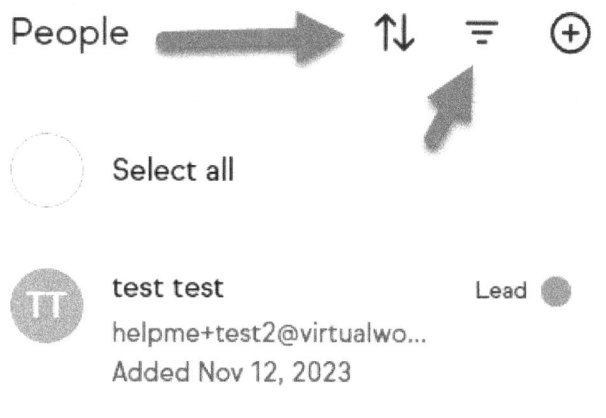

Figure 3.17 – Filtering contacts

3. From the filter drop-down menu, you can select one or more criteria to help narrow down your search. Most criteria can be applied to attributes such as tags, products, names, companies, and so on. Common search attributes include the following:

   A. Contains/does not contain

   B. Is empty/is filled

   C. Equals/not equals

   D. Starts with/ends with

4. Keap will display the number of people in your results to help you get the results you're looking for. To save time in the future, you can save your search criteria by doing the following:

   A. Click the +**Save filtered contacts as a group** link.

   B. Give your group a name in the provided field.

   C. Click the **Save group** button.

5. At any point in your search, you can start over by clicking the **Reset** link and start a new search.

How it works...

Saved searches in a CRM system are like your favorite shortcuts for finding important stuff quickly. Instead of re-enacting a complex search each time, you only need to save it once. So, when you need specific information, such as all your new leads or contacts in the tech industry, you just click on your saved search.

It's a time-saver that keeps you organized and ensures you never miss important details. It's like having a magic button that instantly shows you exactly what you're looking for in your CRM system.

Part 2: Streamlining Communication

This part focuses on communication strategies within Keap, covering email marketing, list management, and personalized messaging to engage and nurture your audience effectively.

- *Chapter 4, Communicating with Your Lists*
- *Chapter 5, Managing Sales Pipelines*

4

Communicating with Your Lists

I'm sure you're all aware that communication is the catalyst that transforms potential leads into valued connections.

Get ready to spice things up with Keap! In this chapter, we'll be jumping right into email marketing. Picture it; the ability to effortlessly schedule and send perfectly seasoned emails, finely tuned to cater to specific interests or groups. You can do that through the power of tagging and list segmentation!

Now, here's the exciting part – you can take it a step further and use emails to gather valuable data, uncovering what makes your audience tick through opens and clicks.

Automation is your trusty sidekick when it comes to communication, making your life easier by delivering targeted messages that not only save time but also crank up engagement and drive conversions.

Keep in mind that every message is a chance to build a stronger connection with your subscribers, turning each interaction into a building block for lasting relationships. Get ready for the next level, where your communication takes center stage and makes a lasting impact!

In this chapter, we'll look at the following recipes:

- Email builder
- Choosing or creating an email template
- Text broadcasts
- Text templates
- Broadcast reports

Technical requirements

For this chapter, you'll require basic copywriting skills.

Email builder

Email broadcasts are like a digital megaphone, letting you speak to a crowd rather than just one person. They're the go-to tool for getting your message out to lots of people all at once.

Learning to use the email builder begins with deciding to use a template or starting from scratch. We're going to examine each step, from implementing drag-and-drop elements such as text, images, and buttons to customizing your design and content. The intuitive interface in Keap makes it easy to create visually appealing and personalized emails without coding expertise.

Whether you're a business owner, a social media guru, or just someone who wants to be heard online, mastering the email builder will take your marketing to the next level.

Getting ready

If you haven't already, you will need to import your contact list(s) from your current marketing tools. We covered this in *Chapter 3*. It is also helpful if you have a basic outline for the email you want to send.

How to do it...

Follow these steps:

1. Click on the **COMMS** tab in the left-hand side navigation bar to open the menu and choose **EMAIL BROADCASTS**.

> **Note:**
>
> Keap will display a list of previously sent and/or draft broadcasts. As you build your email portfolio, this list will be helpful as you can copy an existing email to get you started. For this exercise, we will be starting a brand new email.

2. Click on the **Create email broadcast** button to open the pop-up box:

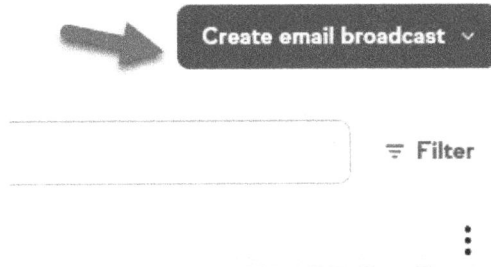

Figure 4.1 – Create email broadcast

3. Select the **Email broadcast NEW** option.

4. By default, the sender will be **The contact's owner**. You can change this by clicking the edit link in the **Sender** section and choosing a different option:

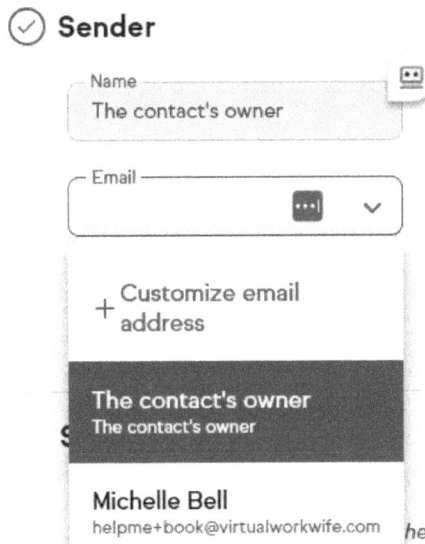

Figure 4.2 – Sender options

5. Choose an existing email from the drop-down menu if you've already set up your email profiles.

Note:

At times, you may want to send a broadcast from an email address that is outside of your default company. This is common when you're running multiple companies through a single CRM. With the updated emailing standards put in place by Gmail, Apple, and Yahoo! in early 2024, this requires you to authenticate any additional domains.

Here are the instructions for authenticating a new domain:

If you haven't set up your profile yet or want to set up a new email, select the **+Customize email address** option.

Click the **Set up email domain** button.

Type the full URL for the domain your email will be sent from (`www.yourdomain.com`)

Choose a host from the menu.

Indicate your **Domain-based Message Authentication, Reporting, & Conformance (DMARC)** status.

Check the box if you have DMARC in place.

If not, a record will be created for you. It's best to choose **none** for security.

Click the **Continue** button.

You will be presented with the data needed to create the appropriate CNAME and TXT records that you will need to create on your web hosting site.

Log in to your domain host and create your records.

You can now return to Keap.

For more information on what email authentication is and how to properly set up your DKIM, DMARC, and SPF, take a look at the following link, where you can watch the email authentication workshop replay: `https://virtualworkwife.com/email-auth-workshop`.

---

6. Click the **Confirm** button when you're done.

7. Click the **Add subject** button.

8. Add your subject in the **Subject line** box (required):

Figure 4.3 – Adding subject content

9. Add some **Preview text**.

> **Note**
>
> Do not leave this blank! Even though it's not required, this is the first impression people get when you show up in their inbox. Use it to your advantage!

10. Click **Save**.

11. Choose your audience by clicking on the **Add audience** button to open the respective menu. You have several options for adding contacts to your broadcast. However, keep in mind that you must have at least *one valid contact* in the audience to send a broadcast:

    A. Search for any tag or saved list by typing the name in the **To** box. You can add a single list or multiple lists and tags.

    B. You have the option to manually add new contacts and include them in your by clicking on the **+Add new contact** option. This is an easy way to add someone quickly to your leads or clients list so that they are included in your broadcast:

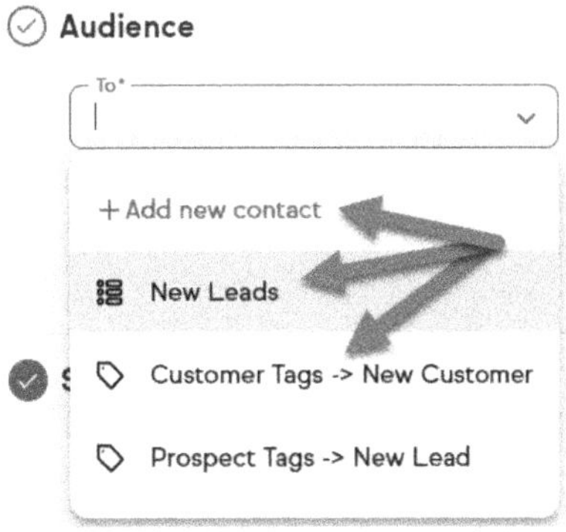

Figure 4.4 – Audience options

12. Click **SAVE**.

13. Set the time and date you want to send your email by editing the **Schedule** option:

A. Select **Right away** if you want your broadcast to be sent immediately.

B. Choose **At a specific time** if you plan to send your email in the future.

C. Selecting to send using **Contact's time zone** requires carefully considering the exact time you plan to send the email as it may occur "in the past," depending on the difference between your time zone and your recipient's:

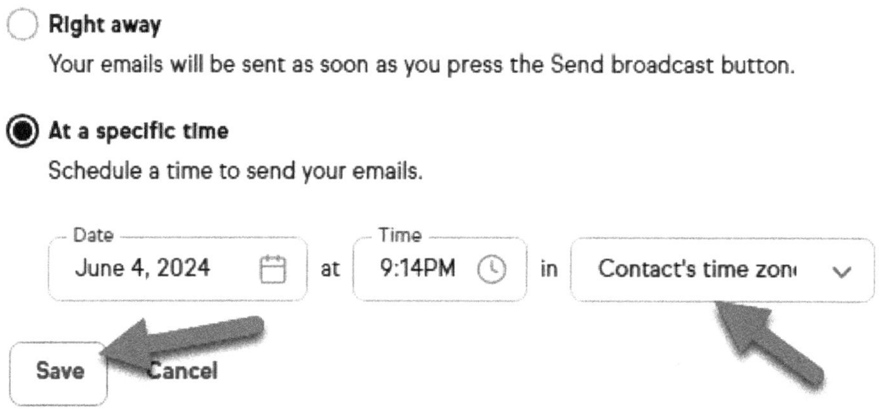

○ **Right away**
Your emails will be sent as soon as you press the Send broadcast button.

◉ **At a specific time**
Schedule a time to send your emails.

Date		Time		in	Contact's time zon ⌄
June 4, 2024 📅	at	9:14PM 🕐			

Save   Cancel

Figure 4.5 – Setting your send time

How it works...

By setting **From**, **To**, **Subject**, and **Preview text**, as well as a thoughtful schedule, you're setting up a system that's both efficient and scalable for sharing important information, announcements, or promotions. Email broadcasts are a go-to for anyone looking to connect with their audience, and using techniques such as attention-grabbing subject lines and personalized content to boost open rates will ensure a lasting impact. Digging into the analytics of each broadcast will help you tweak your approach, making sure your email automations hit your audience when they're most likely to engage.

Choosing or creating an email template

You are now ready to create your content. Each time you send a broadcast, you create an email template that has long-term value beyond its initial use.

Creating email templates holds significant value as it streamlines and standardizes the communication process, saving time and ensuring consistency in messaging. They enable quick customization, allowing users to adapt content while maintaining a branded and polished appearance. Templates not only enhance efficiency but also contribute to building a recognizable and trustworthy email presence, making them a valuable asset for anyone looking to communicate effectively.

In this recipe, we'll go through the steps of creating email templates to streamline communication, save time, ensure consistency, and build a recognizable brand presence.

Getting ready

In most cases, you will be using standard fields (name, email company, and so on) to populate information in your email communications. It is possible to capture more personal information, something we covered in *Chapter 3*, by using custom fields. That information can also be inserted into your email communications, making automated messaging much more personal. For this recipe, we'll stick to using standard fields.

How to do it...

The email builder is broken into five sections:

1. Content
2. Blocks
3. Body
4. Images
5. Uploads

We will discuss each section and how they work together to create a fun and functional tool for you to communicate with your list.

Creating content

This is the largest of the five sections and where you'll spend a lot of your building time. There are 14 tools in the content builder to help you create dynamic, educational, and/or entertaining emails to serve your audience. Follow these steps:

1. To continue, click **Create content** in the right-hand side menu bar:

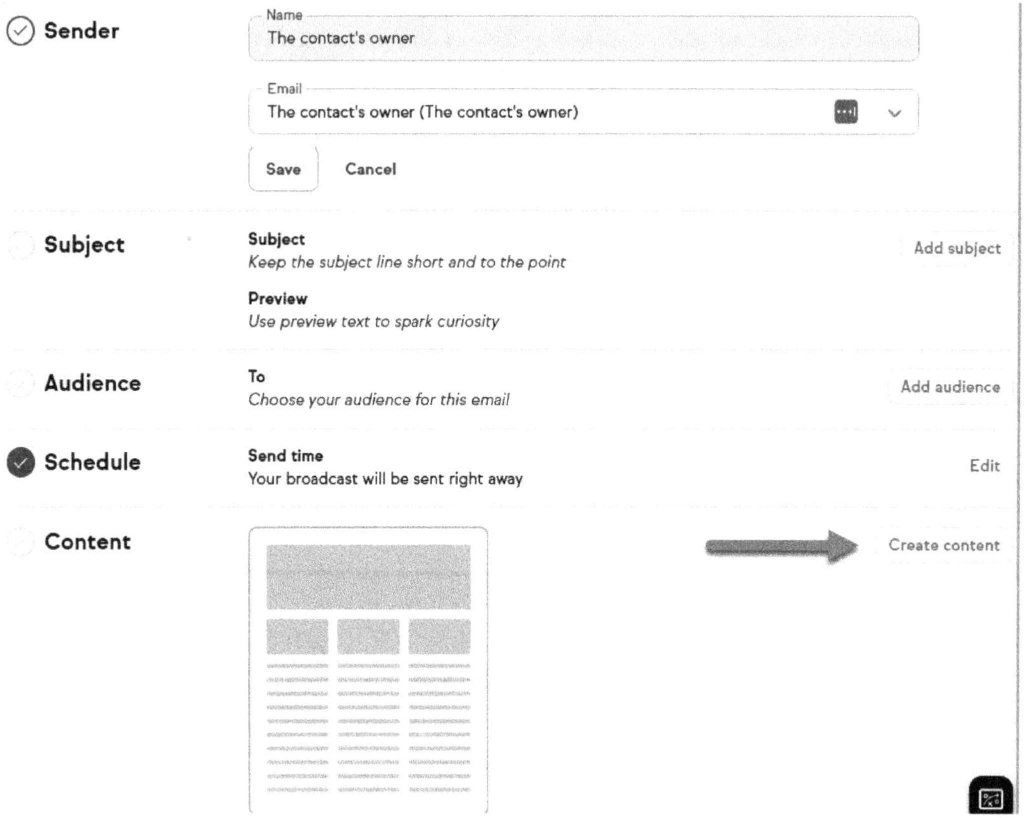

Figure 4.6 – Clicking Create content

2. You now have the option to do the following things:

    A. Start from scratch and design your own email using HTML or plain text.

    B. Choose a template from the **Gallery** area.

    C. Choose from your own saved templates.

    D. Choose to copy a previously sent email.

    E. Continue a previously started draft.

For this recipe, we will be selecting **Start from scratch**:

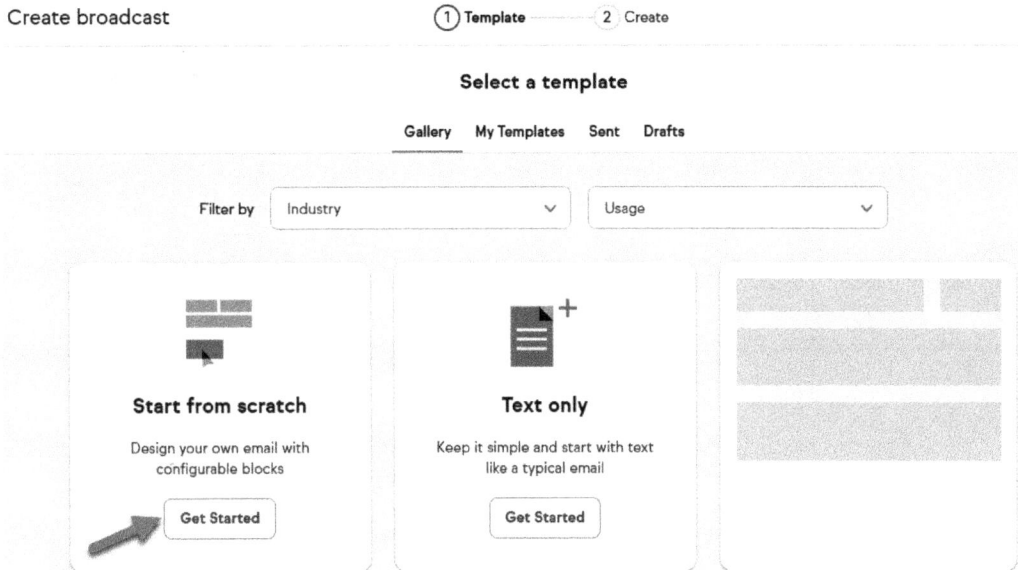

Figure 4.7 – Adding email content

3. Click the **Get Started** button in the **Start from scratch** box to open the email editor.

> **Note**
>
> The email builder is a drag-and-drop tool that allows you to easily add features to your email template. By default, the editor starts in desktop view, meaning what you're seeing is what someone reading your email on a computer would see. You can change the view between desktop and mobile by using the toggle at the top of the editor. It is highly recommended that you switch to the mobile view when creating your template to ensure that what a mobile reader sees is formatted for optimal reading.

Now that we're in the email builder, let's remember that there are 14 tools we'll be working with. Each tool plays a significant role in how people see and consume your email content. You will move between the canvas and the right-hand side tool menu as you work with your template.

The canvas is made up of rows or sections, and columns. When you're working with a template, you can apply design rules to both the section (row) and column:

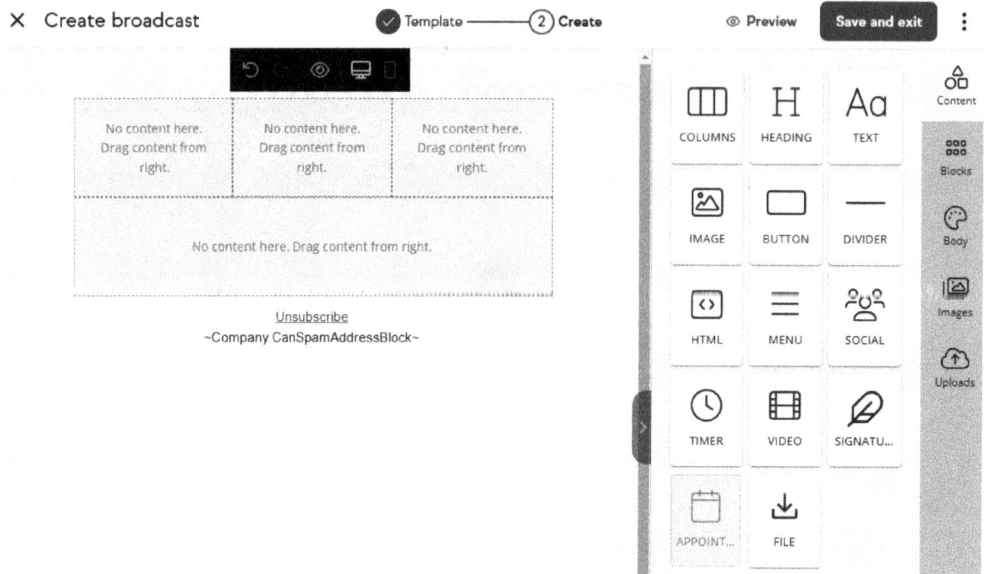

Figure 4.8 – Email builder tools

Columns

Columns are what help you style your content in a visually pleasing way. When working in the Keap email builder, which is mobile-responsive, think of columns as your trusty tool for keeping the content of your section (row) tidy. Whether you're going for a single-column layout or getting a bit more creative with up to three columns, as shown in *Figure 4.7*, columns give you the power to showcase content or images side by side. To add columns to your email template, follow these steps:

1. Begin by clicking on and then dragging the **Columns** icon from the **Tools** menu onto the email canvas.

2. There are six column layout options for you to choose from. From the Tools menu, click on the option you prefer.

3. Set a background color or image if desired.

4. Set padding to **push** your content away from the edges of the column.

5. Columns work best in desktop view. When switching to mobile view, you may notice that your column content is not aligned the way you expect. You have the option to hide or force the columns to not stack by adjusting the toggle under **RESPONSIVE DESIGN**:

    A. On the toolbar, click the **Mobile** tab.

B. Scroll to the bottom of the toolbar and use the toggle to engage the **Do Not Stack on Mobile** option:

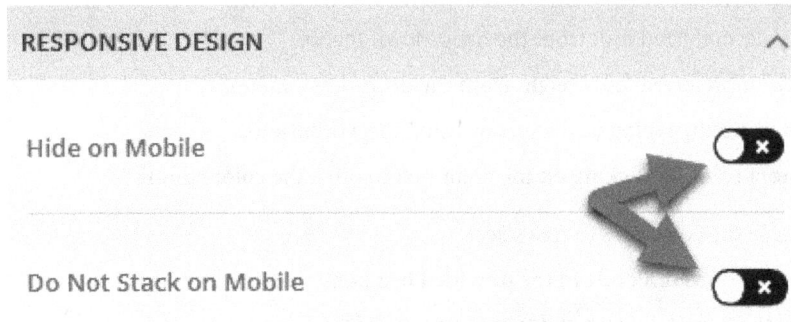

Figure 4.9 – Toggle buttons

Headings

Headings in an email template serve as the titles or main sections that structure the content and guide the reader through the message. These headings are typically larger, bolded, or stylized text that stands out, making it easy for recipients to identify and navigate different parts of the email. Effective headings in an email template can enhance readability, drawing attention to key information.

To add a heading to your email, follow these steps:

1. Begin by dragging the **Header** icon from the **Tools** menu onto the email canvas.

2. Choose your heading size by clicking on **H1, H2, H3,** or **H4**.

3. Choose your preferred font from the drop-down menu.

4. Choose your preferred font weight from the drop-down menu.

5. Set the font size by typing in the box or using the +/- buttons.

6. Align your header by clicking on the left, center, right, or justified button.

7. You can set the line height by typing in a percentage or using the +/- buttons.

8. Use the **Responsive Design** toggles to hide or display your heading on mobile and desktop devices.

Text

The text in your email template should be clear, concise, and tailored to the intended audience. It may consist of various elements, such as introductory greetings, main message content, calls to action, and closing remarks. Proper formatting, including font styles, sizes, and colors, is essential to ensure readability and visual appeal. Well-crafted text is crucial for effective communication, allowing the sender to convey their message persuasively and engagingly to the email recipient.

To add text to your email, follow these steps:

1. Begin by dragging the **Text** icon from the **Tools** menu onto the email canvas.

2. Choose your preferred font from the drop-down menu.

3. Choose your preferred font weight from the drop-down menu.

4. Set the font size by typing in the box or using the +/- buttons.

5. Select a font color by clicking on the color box to open the color editor:

   A. Click in the color area to free select.

   B. Type in your HEX code in the provided hex box.

   C. Type in your RGB color codes in the provided RGB boxes.

6. Align your text by clicking on the left, center, right, or justified button.

7. You can set the line height by typing in a percentage or using the +/- buttons.

8. Use the **Responsive Design** toggles to hide or display your heading on mobile and desktop devices.

Images

Images in an email serve as visual elements that complement or enhance the text content, providing a more engaging and impactful communication experience. These can include photos, graphs, logos, icons, or any visual representation embedded within the email body. Images are often used to convey information, illustrate key points, or evoke emotions. Alt text, or alternative text, is essential for images in emails, providing a brief description of the image content for recipients who may have images disabled or are using screen readers. Including relevant and compelling images in an email can capture the recipient's attention, reinforce branding, and create better conversion.

To add an image to your email, do the following:

1. Begin by dragging the **Image** icon from the **Tools** menu onto the email canvas.

2. There are several ways to select an image for your email:

   A. Search previously uploaded images

   B. Choose from stock photos.

   C. Drop new images into the box.

   D. Insert the email using a URL.

3. You can set the width of your image manually but toggling the **Auto Width** button off.

4. Align your image by clicking on the left, center, right, or justified button.

5. Add your alternative text in the provided box.

6. Set an action for your image by selecting one from the drop-down menu:

   A. **Open Website**:

      i. Add the URL for the website in the URL box.

   B. **Send Email**:

      i. Add the mail to address in the **Mail to** box.

      ii. Add your subject to the **Subject** box.

      iii. Add any body text to the **Body** box.

   C. **Call Phone Number**:

      i. Add the number you want to be dialed to the **Phone** box.

7. Use the **Responsive Design** toggles to hide or display your text on mobile and desktop devices.

Buttons

Buttons in an email are interactive elements that are designed to prompt a specific action from the recipient when clicked. They are your "call to action" and encourage people to click through to your website, download a document, or make a purchase. Buttons should be stylized to stand out, using contrasting colors, or prominent placement. Effective button design involves clear and concise text on the button itself, indicating the action it performs, and ensuring that it is visually appealing and easily clickable. Including buttons in an email enhances the user experience, providing a straightforward and visually intuitive way for recipients to respond to the sender's message.

To add a button to your email, follow these steps:

1. Begin by dragging **Button** onto the email canvas.

2. Set an action for your button by selecting one from the drop-down menu:

   A. **Open Website**: Send the reader to a URL you specify:

      i. Add the URL for the website in the URL box.

   B. **Send Email**: Opens the reader's email tool and initiates an email with your information:

      i. Add the mail to address in the **Mail to** box.

      ii. Add your subject to the **Subject** box.

      iii. Add any body text to the **Body** box.

   C. **Call Phone Number**: Opens the reader's default phone dialer and calls the number you define:

      i. Add the number you want to be dialed to the **Phone** box.

3. Set your button's text color and background color by clicking on the respective boxes. Then, do the following:

   A. Click in the color area to free select.

   B. Type in your HEX code in the provided hex box.

   C. Type in your RGB color codes in the provided RGB boxes.

4. Use the design toggle to turn **Auto Width** on/off:

   A. **On**: The button will automatically size to its container

   B. **Off**: Uses the slider to set the button percentage for the container

5. Choose your preferred font from the drop-down menu.

6. Choose your preferred font weight from the drop-down menu.

7. Set the font size by typing in the box or using the +/- buttons.

8. Align your text by clicking on the left, center, right, or justified button.

9. You can set the line height by typing in a percentage or using the +/- buttons.

10. Button padding can be set to the same for all sides or customized on each side by using the toggle button.

11. Button borders:

    A. Set the button's border text color and background color by clicking on the respective boxes. Then, do the following:

       i.   Click in the color area to free select.

       ii. Type in your HEX code in the provided hex box.

       iii. Type in your RGB color codes in the provided RGB boxes.

    B. The button border can be set to the same for all sides or customized on each side by using the toggle button.

12. Use the **Responsive Design** toggles to hide or display your button on mobile and desktop devices.

Dividers

Dividers are horizontal lines that are used to visually separate different sections of content within the email template. These dividers help create a clear and organized layout, preventing the email from appearing cluttered and making it easier for recipients to navigate the information. By breaking up the content into distinct sections, dividers contribute to a more aesthetically pleasing and user-friendly email design. They are especially useful in longer emails or newsletters.

To add a divider to your email, follow these steps:

1. Begin by dragging the **Divider** icon onto the email canvas.

2. Use the slider to set the width of your divider.

3. Use the drop-down menu to choose your line type:

    A. Solid _____

    B. Dotted ....

    C. Dashed ------------

4. Align your divider by clicking on the left, center, right, or justified button.

5. Use the **Responsive Design** toggles to hide or display your heading on mobile and desktop devices.

HTML

HTML is commonly used to design and customize templates when the onboard tools don't provide enough customization to suit your needs.

In emails, HTML serves a similar purpose as it does on websites, but with some differences. Instead of creating a whole web page, HTML in emails is used to structure and style the content of the email itself.

For instance, you can use HTML to format text (such as making some words bold or italic), insert images, create links, and even add basic layout elements such as tables. HTML allows you to design visually appealing emails and make them more engaging for recipients.

However, it's important to note that not all email clients render HTML in the same way, and some users may have email settings that disable HTML rendering for security reasons.

To add HTML to your email, do the following:

1. Begin by dragging the **HTML** icon onto the email canvas.

2. Paste your HTML code into the provided box.

3. Use the **Responsive Design** toggles to hide or display your HTML on mobile and desktop devices.

Menu

A menu in an email typically refers to navigation that allows the recipient quick clickable access to different sections within the email content. This is often seen in newsletters or marketing emails where the sender wants to direct recipients to specific areas, such as featured products, articles, or landing pages. Including a menu in an email enhances user experience, making it easy for recipients to find and engage with the content that interests them the most.

To add a menu to your email, follow these steps:

1. Begin by dragging the **Menu** icon onto the email canvas.

2. From the toolbar, click +**Add New Item** to open the editor and begin adding your items.

3. In the **Text** box, type what your menu item should say.

4. In the action dropdown, choose your desired outcome:

    A. **Open Website**:

        i. Add the URL for the website you want them to visit in the URL box.

        ii. You can apply a tag when someone clicks the menu item by selecting it from the drop-down menu.

    B. **Send Email**:

        i. Add the mail to address in the **Mail to** box.

        ii. Add your subject to the **Subject** box.

        iii. Add any body text to the **Body** box.

    C. **Call Phone Number**:

        i. Add the number you want to be dialed to the **Phone** box.

5. Repeat *Steps B* to *C* until you have added all your menu options.

6. Choose your preferred font from the drop-down menu.

7. Choose your preferred font weight from the drop-down menu.

8. Set the font size by typing in the box or using the +/- buttons.

9. Select a text color by clicking on the color box to open the color editor.

    A. Click in the color area to free select.

    B. Type in your HEX code in the provided hex box.

    C. Type in your RGB color codes in the provided RGB boxes.

10. Select a link color using the same method used in *Step 8*.

11. Align your text by clicking on the left, center, right, or justified button.

12. Choose a horizontal or vertical layout from the drop-down box.

13. You can add a separator between menu items by inserting a symbol (| or :) in the separator box.

14. Use the **Responsive Design** toggles to hide or display your menu on mobile and desktop devices.

Socials

Social icons in an email allow recipients to connect with you or share email content on social networks. These icons typically include well-known platforms such as Facebook, Twitter, Instagram, LinkedIn, and others. Placed strategically in the email layout, social icons encourage recipients to engage with your social media profiles, helping to grow your authority.

To add social icons to your email, do the following:

1. Begin by dragging the **Socials** icon onto the email canvas.
2. There are three types of icons to choose from with three color options. From the toolbar, click the **Icon Type** dropdown and select the following options:

    A. **Circle**: Full color.

    B. **Circle Black**: Black background with white image.

    C. **Circle White**: White background with a gray image.

    D. **Round (square with rounded edges)**: Full color.

    E. **Round (square with rounded edges)**: Black background with white image.

    F. **Round (square with rounded edges)**: White background with gray image.

    G. **Square**: Full color.

    H. **Square**: Black background with white image.

    I. **Square**: White background with a gray image.

3. Next, click on the social icon you want to include.
4. Add your URL in the provided space.
5. Repeat *Steps 3* to *4* to add your desired social channels.
6. Align your icons by clicking on the left, center, right, or justified buttons.
7. Set icon spacing by typing in the box or using the +/- buttons.
8. Use the **Responsive Design** toggles to hide or display your socials on mobile and desktop devices.

Timers

Dynamic countdown timers are embedded in email content to display the remaining time for a specific offer, promotion, or event. These timers are often used in marketing emails to create a sense of urgency and encourage recipients to take immediate action. The countdown may be linked to a limited-time discount, a product launch, or an event registration deadline. By leveraging timers, you can capture the attention of your recipients, driving engagement, and prompting them to make quicker decisions.

To add a timer to your email, follow these steps:

1. Begin by dragging the Timer icon onto the email canvas.

2. On the toolbar, a default time will be displayed in the **End Time** box. To choose your **End Time** value, click on the box to open the calendar popup:

    A. Use the <> buttons to navigate to the month and year and select your date.

    B. Choose a time for your offer to expire on that date.

3. In the **Timezone** box, select your preferred timezone.

4. In the **Language** box, select the preferred language for your timer display.

5. Select your background, digits, and label colors by clicking on the appropriate color box to open the color editor:

    A. Click in the color area to free select.

    B. Type in your HEX code in the provided hex box.

    C. Type in your RGB color codes in the provided RGB boxes.

6. Choose your preferred digit font from the drop-down menu.

7. Choose your preferred label font from the drop-down menu.

8. You can set the width of your timer manually by toggling the **Auto Width** button off.

9. Align your timer by clicking on the left, center, right, or justified button.

10. Add **Alternate Text** to the box.

11. You can set an action if someone clicks on the timer by choosing an option from the image link drop-down menu:

    A. **Open Website**: Send the reader to a URL you specify:

        i.   Add the URL for the website in the URL box.

        ii. You can apply a tag when someone clicks the menu item by selecting it from the drop-down menu.

    B. **Send Email**: Opens the reader's email tool and initiates an email with your information:

        i.   Add the mail to address in the **Mail to** box.

        ii. Add your subject to the **Subject** box.

        iii. Add any body text to the **Body** box.

    C. **Call Phone Number**: Opens the reader's default phone dialer and calls the number you define:

        i.   Add the number you want to be dialed to the **Phone** box.

12. Use the **Responsive Design** toggles to hide or display your timer on mobile and desktop devices.

Videos

Embedding videos in emails involves incorporating playable video content directly into the email message, allowing recipients to view the video without leaving their email client. This can be achieved using HTML5 video tags or, more commonly, by including an image or a thumbnail linked to an external video hosting platform, such as YouTube or Vimeo. When recipients click on the video thumbnail or link, they are redirected to a web page or a landing page where the video can be played.

However, it's important to note that not all email clients support embedded video playback, and certain security or privacy settings might prevent videos from playing directly within the email.

As a result, many email marketers use enticing video thumbnails or GIFs with a "play" button, prompting users to click through to an external platform to watch the video. Despite these limitations, including video in emails can be a powerful way to engage recipients with dynamic and visually appealing content.

To add a video to your email, do the following:

1. Begin by dragging the **Video** icon onto the email canvas.
2. Add your YouTube or Vimeo URL to the URL box.

> **Important note**
> The **Video** icon only works with YouTube and Vimeo.

3. Set padding to **push** your video away from the edges of the email.
4. Use the **Responsive Design** toggles to hide or display your video on mobile and desktop devices.

Signatures

In *Chapter 2*, we discussed creating your personal avatar. The information you added to your profile laid the groundwork for us to be able to add your signature to your email templates.

A signature in an email is a personalized block of text, often located at the end of the email, that provides information about the sender. Typically, an email signature includes your name, title, company, or affiliation, contact information, and sometimes a logo or other relevant details.

Email signatures serve as a professional and consistent way to conclude an email, offering recipients additional context about the sender and making it easy for them to get in touch with you. Using email signatures is a great way to reinforce your brand identity and provide a polished and cohesive appearance to your email.

To add a signature to your email, follow these steps:

1. Begin by dragging the **Signature** icon onto the email canvas.

2. You have the option to pick any user, and their signature will appear in the email body. Alternatively, you can select **contact owner** to automatically send the email from the user who is linked to the recipient.

3. Set padding to **push** your signature away from the edges of the email.

4. Use the **Responsive Design** toggles to hide or display your signature on mobile and desktop devices.

Appointments

Adding clickable buttons that allow people to easily book a specific type of call with you is an invaluable tool when it comes to email marketing. This tactic is commonly used in professional communication to simplify the process of coordinating schedules. Instead of engaging in back-and-forth emails to find a suitable time, the sender includes a scheduling link that directs the recipient to a calendar interface, displaying open slots for appointments. Recipients can then choose a convenient time without the need for extensive coordination. Keap's built-in scheduling tool means you don't have to pay for an additional resource such as Calendly or Oncehub.

To add an appointment button to your email, do the following:

1. If you haven't already, you will need to follow the instructions provided in *Chapter 2* for setting up the appointment type. If you don't have an active appointment type, you won't be able to drag the **Appointment** icon into your email.

2. Begin by dragging the **Appointment** icon onto the email canvas.

3. From the drop-down menu, choose the appointment type you want to include in your email.

4. Add the `call to action` text for your appointment button.

5. Select your background and text color by clicking on the appropriate color box to open the color editor:

    A. Click in the color area to free select.

    B. Type in your HEX code in the provided hex box.

    C. Type in your RGB color codes in the provided RGB boxes.

6. Set the width of your button manually by toggling the **Auto Width** button off.

7. Set the line height by typing in a percentage or using the +/- buttons.

8. Choose your preferred font size by typing in a value or using the +/- buttons.

9. Choose a font style from the **Font Family** dropdown.

10. Choose your preferred font from the drop-down menu.

11. Use the **show more options** link to open additional formatting options such as padding, margins, border style, and margins.

12. Use the **Responsive Design** toggles to hide or display your signature on mobile and desktop devices.

Files

You can attach files to an email broadcast. These attachments can be any type of file, such as Word documents, PDFs, images, spreadsheets, or multimedia files. When a sender attaches a file to an email, the recipient can download and open the attached file on their device. However, it's important to be mindful of file size limits imposed by email providers. Large attachments may be rejected or result in issues with email delivery.

To attach a file to your email, follow these steps:

1. Begin by dragging the **File** icon onto the email canvas.

2. Click the **Open File Manager** button to open the pop-up menu.

3. To attach a previously uploaded file, you must do the following:

   A. Scroll the list of available files and click on the one you want. Note that having a standard naming convention for your files makes it a lot easier to locate them in the future.

   B. Click the **Insert File** button to continue.

4. To upload a new file, follow these steps:

   A. Begin by dragging the **File** icon onto the email canvas.

   B. Click the **Open File Manager** button to open the pop-up menu:

      i. Drag and drop a file into the box or use the **Browse** link to search your PC.

5. Once uploaded, your file will be added to the file list. Simply select the file and click the **Insert File** button to continue.

6. You can change the name of the **Download File** button by editing the text in the tools menu.

7. Set your button's text color and background color by clicking on the respective boxes:

   A. Click in the color area to free select.

   B. Type in your HEX code in the provided hex box.

   C. Type in your RGB color codes in the provided RGB boxes.

8. Align your button by clicking on the left, center, right, or justified button.

9. Use the design toggle to turn **Auto Width** on/off:

    A. **On**: The button will automatically size to its container.

    B. **Off**: Uses the slider to set the button percentage for the container.

10. Set the line height by typing in a percentage or using the +/- buttons.

11. Set the font size by typing in the box or using the +/- buttons.

12. Choose your preferred font from the drop-down menu.

13. Choose your preferred font weight from the drop-down menu.

14. Use the **Responsive Design** toggles to hide or display your button on mobile and desktop devices.

Attaching blocks to your email

Blocks are pre-designed and customizable content elements or sections that you can drag and drop into your email layout to create a visually appealing and structured design. This will save you a lot of time and effort when designing your emails. These blocks make it easy for users with varying design skills to create professional-looking emails without having to code from scratch. Once you've added a block to your email, you can arrange, customize, and edit it by going back to the content section. Blocks help you streamline email creation, offering a convenient way for you to craft visually engaging and cohesive email campaigns.

To attach a block to your email, follow these steps:

1. To begin, click the **Blocks** icon on the right-hand side menu bar.

2. You now have the option to do one of the following:

    A. Search for a predesigned block by style (holiday, event, and so on).

    B. Add a blank block.

3. Clicking the **all** link next to a style will filter the list for all blocks similar to that style.

4. Continue by dragging your selected block onto the email canvas.

The content editor will automatically open so that you can customize your block. To add more blocks, simply close the content editor by clicking the **X** icon in the top-right corner of the menu bar.

Body

Establishing general settings for your email is advantageous as it ensures consistency in your design and streamlines the email creation process. Brand identity plays an important role in all your communications, especially email marketing. General settings help you create a cohesive image and reduce the risk of errors, which enhances the overall user experience.

> **Note**
>
> If you select a color for the email's background, it will span the entire email. If the columns within the email are not assigned a color, such as white, they will be transparent, allowing the background color or image to show through.

1. To begin, click the **Body** icon on the right-hand side menu bar.

2. Set your default text and background color by clicking on the respective boxes:

    A. Click in the color area to free select.

    B. Type in your HEX code in the provided hex box.

    C. Type in your RGB color codes in the provided RGB boxes.

3. Set your content width by typing in the box or using the +/- buttons.

4. Set your alignment by clicking on the left, center, right, or justified button.

5. Choose your preferred font from the drop-down menu.

6. Choose your preferred font weight from the drop-down menu.

7. Set your default link color by clicking on the respective boxes and doing the following:

    A. Click in the color area to free select.

    B. Type in your HEX code in the provided hex box.

    C. Type in your RGB color codes in the provided RGB boxes.

8. You can turn the link underling on/off by moving the toggle button left/right.

9. Continue by dragging your selected block onto the email canvas.

These settings create a foundation that not only saves time but also contributes to a polished and recognizable brand image, making your emails more efficient, error-resistant, and professional.

Adding images

The **Images** section allows you to search millions of photos from several sources using one tool. All images are licensed under Creative Commons Zero. Follow these steps:

1. To begin, click the **Images** icon on the right-hand side menu bar.

2. Enter search criteria into the provided box.

3. To add an image to your design, drag it over the email canvas and place it where you want it to be displayed.

The content editor will automatically open so that you can customize your image. To add more images, simply close the content editor by clicking the **X** icon in the top-right corner of the menu bar.

Uploading images

The **Uploads** section allows you to retrieve any images you have previously selected from the library or uploaded from your files. Follow these steps:

1. To begin, click the **Uploads** icon on the right-hand side menu bar.

2. To upload a new image, do the following:

    A. Click the **Upload Image** button.

    B. Drag an image from your files over the box and drop it.

3. Navigate to your image in the gallery and drag it onto the email canvas.

Merge fields

Merge fields in a CRM are placeholders that dynamically pull personal data from the contact record into emails, forms, texts, or other communication materials. These fields enable you to create tailored and individualized content at scale, addressing recipients by their names, incorporating specific details, or customizing messages based on unique attributes.

For example, in an email campaign, merge fields can be utilized to automatically insert a contact's name, company, or any other relevant information, providing a more personal and engaging experience.

Let's begin by navigating to the **Content** section of our email builder. You can do this by clicking on the **Content** icon on the right-hand side menu bar. From here, do the following:

1. Drag the **Text** icon onto the email canvas.

2. Click inside the text box. You want to put your cursor where you want to add your merge field. Once you click in the box, the editing tool will pop up:

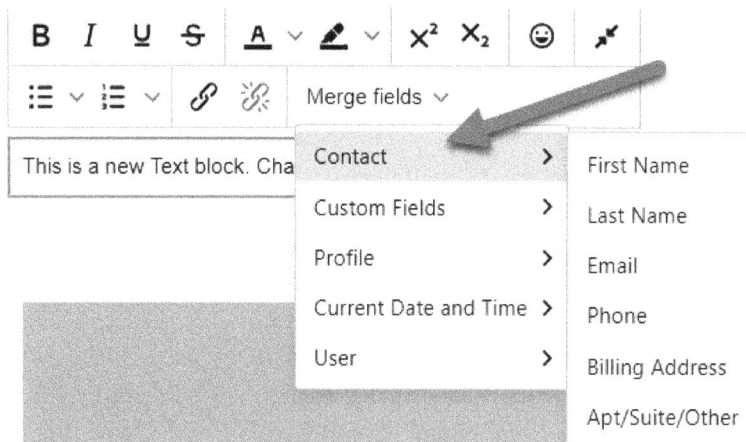

Figure 4.10 – Viewing merge fields

3. Click **Merge fields** to open the options menu. From here, you have the following options:

    A. **Contact**: Basic contact information such as address, phone, email, and more

    B. **Custom fields**: Additional data fields that you create and define (see *Chapter 3*)

    C. **Profile**: These are fields about *you* – your company information, logo, and so on

    D. **Current Date and Time**: Easy date and time options

    E. **User**: Your name and email

4. Click on the field you want to add to your email. The field will automatically be added where your cursor is – for example, `{{profile.company.name}}`.

5. Continue to edit your text as needed, adding merge fields where appropriate.

Tying it all together

So, you've selected your audience, added your content, formatted your blocks, uploaded your images, and used merge fields to create a highly personal and polished email broadcast. There are just a couple more steps and you'll be well on your way to email marketing genius!

It's time for the *most* important step: test, test, test!

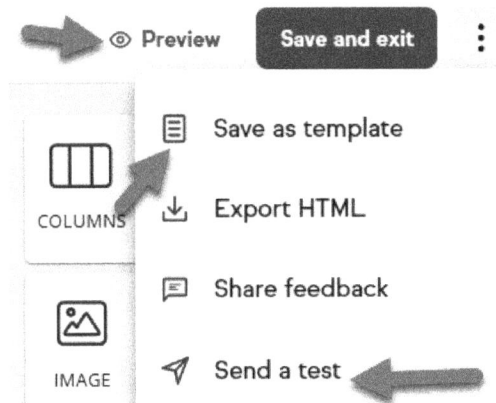

Figure 4.11 – Testing and sending an email

Follow these steps:

1. Click on the eye to preview your test and be sure to look at both mobile and desktop versions.

2. If a section doesn't look good on both mobile and desktop, do the following:

    A. Click on the item.

B.   Click the **Duplicate** button in the top-right corner of the menu bar:

Figure 4.12 – The Duplicate button

C.   Set the **Responsive Design** toggle to on for desktop; do the same for mobile.

D.   Edit each item and preview them until you like the results.

3.   Now, you must send yourself a test:

A.   Click the three black dots to open the menu.

B.   Click the **Send a test** option.

C.   Choose your email from the drop-down list.

D.   Click the **Send Test Email** button.

E.   Open your email client (Google, Yahoo!, and so on).

F.   Look for the email:

   i.   Did it go to spam?

   ii. If so, consider editing your subject line.

   iii. Have you authenticated your email address?

G.   Open the email and click on every image. Does it do what you expected?

   i.   If not, edit and retest.

H.   Click on every link. Does it navigate to the right location?

   i.   If not, edit and retest.

4.   After extensive testing, your email is ready to use.

5.   Navigate back to your Keap application to continue.

6.   You may want to save your new email as a template before sending it. This can save you time later on. Saving a copy of your latest webinar invitation as a template can save you time when you're ready to launch your next webinar. To save your email as a template, follow these steps:

A.   Click the three black dots to open the menu.

B.   Click the **Save as a Template** icon.

C. Give your template a name.

D. Click the **SAVE** button.

7. To send your email, do the following:

A. Click the **SAVE and EXIT** button to return to the builder.

B. Click the **Send Broadcast** button to send your email.

How it works...

Creating an email template involves several key steps – from identifying your recipients to structuring your message with headings, text, images, and buttons all while keeping it concise and engaging. By following these steps, you can enhance reach and engagement, allowing businesses and individuals to connect with you on a more personal level. With strategic use of elements such as catchy subject lines and targeted content, broadcast emails can captivate and influence your potential leads and clients. Overall, broadcast emails serve as a scalable and impactful tool for information sharing, marketing campaigns, fostering connections, and creating automations to help you reach more people faster.

Text broadcasts

Text message broadcasts complement and support your email broadcasts by offering an additional channel for reaching your audience. While email broadcasts provide a comprehensive platform for detailed and visually rich communication, text messages offer immediacy and directness.

By combining both channels, you can create a multi-faceted communication strategy that caters to the diverse preferences and habits of your audience. Text messages are often more accessible and get quicker responses, making them ideal for time-sensitive announcements or urgent promotions.

The synergy between email and text message broadcasts ensures a broader reach, increased engagement, and a more effective overall communication strategy for your business or personal brand.

Getting ready

To send text broadcasts responsibly, it's important to make sure you have a clean phone list of *mobile* numbers and that you've obtained the recipients' permission to send texts. You'll want to craft concise yet engaging messages, include an opt-out mechanism for those who want to get off your list, and follow relevant regulations.

If you haven't done so already, you will need to set up your Keap marketing number to continue. Refer to *Chapter 2*, the *Obtaining your phone numbers* section, for more details.

How to do it...

Follow these steps:

1. Click on the **COMMS** tab from the left-hand side navigation bar to open the menu. Choose **TEXT MESSAGE BROADCASTS**.

    Keap will display a list of previously sent and/or draft broadcasts. As you build your text portfolio, this list will be helpful as you can copy an existing email to get you started. For this recipe, we will be starting a brand new text:

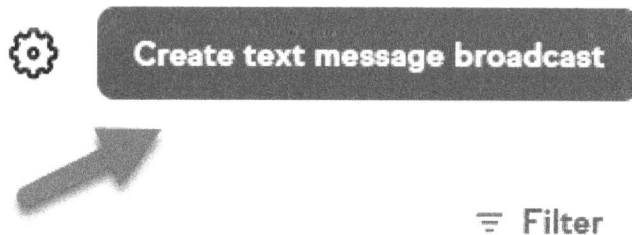

    Figure 4.13 – Create text message broadcast

2. Click on the **Create text message broadcast** button to open the pop-up box.
3. Name your broadcast.
4. Your Keap phone number will already be plugged into the **from** box.
5. Choose your audience by clicking on the audience box. You can send texts to the following:

    A. Saved searches (groups).

    B. Tags.

    C. Individual contacts.

6. Write your message in the text box. As you write, you can preview your email in the adjoining window, as shown in *Figure 4.12*:

Preview

This is my cool text i'm sending to [[contact.first_name]]! See you on zoom tomorrow. here's the link: www.zoom.com

(833) 437-0645 Just now

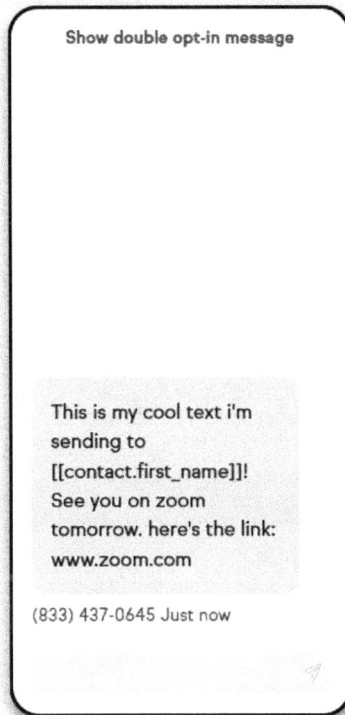

Figure 4.14 – Text preview

7. Click the # symbol to insert merge fields, such as contact name or the date/time.

    160 characters or less is considered "one" text. Larger messages will count as multiple texts.

8. Choose what type of phone number you want to include in your broadcast.

> **Note**
> Choose a "best practice" that works for you, such as always storing mobile phones in the **phone 3** field.

9. Click the **Review Broadcast** button.

10. A certification box will pop up, asking you to confirm that you have permission to text the list you compiled. Read it carefully and then proceed by ✓checking the box and clicking **Continue**.

11. At this point, you'll be on the **Review and Send** page. From here, do the following:

    A. Confirm your recipients are correct.

    B. Ensure an estimated price for the text will be displayed.

    C. Confirm your message content:

        i. Is the tone correct?

        ii. Is everything spelled correctly?

        iii. Do you have a call to action?

        iv. Click the **X** icon in the top-left corner to go back and make edits if needed.

12. Now is the perfect time to send yourself a test so that you can see what it looks like from the end user's perspective.

13. If you're ready, you can proceed and send your broadcast:

    A. Choosing **Now** will send your text immediately.

    B. Choosing **Later** will allow you to set the date, time, and time zone for when you want to send your message in the future.

14. Click the **Schedule Broadcast** button to complete your text.

15. A final warning message will appear, giving you one last chance to make edits. If you're ready, choose **I understand, send broadcast**.

How it works...

Consider the recipients' time zones and conduct thorough testing before sending your broadcast text messages. With these essential elements in place, you'll be well-prepared to implement a successful and compliant text broadcasting strategy. This approach not only enhances your marketing conversion rates but also accelerates your time-to-close metrics, ensuring a more effective and impactful outreach to your audience.

Text templates

Creating templates for text broadcasting offers significant value by streamlining the process of crafting and sending messages to a wider audience. Templates provide a standardized format, ensuring consistency and professionalism in your communications. They save time by eliminating the need to recreate messages from scratch and allow for quick customization. Additionally, templates contribute to brand consistency and message cohesion, reinforcing your brand identity.

How to do it...

Follow these steps:

1. Click on the **COMMS** tab from the left-hand side navigation bar to open the menu. Choose **TEXT MESSAGE BROADCASTS**.

   Keap will display a list of previously sent and/or draft broadcasts. As you build your text portfolio, this list will be helpful as you can copy an existing email to get you started. For this recipe, we will be starting a brand new text:

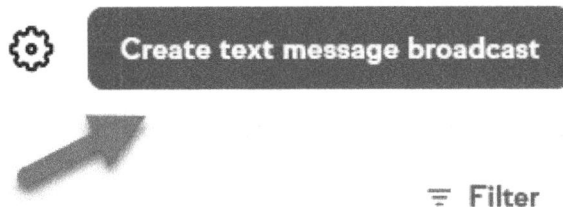

Figure 4.15 – Create text message broadcast

2. At this point, since we want to create a template and not send a text, we're going to skip over several of the steps (because we just want to save a template that is inside the next step) and navigate directly to the **write your message** box:

   A. Click on the ▤ icon in the lower-left corner of the box.

   B. A list of existing templates will appear.

   C. Click on the **Manage Templates** link at the bottom of the popup.

3. An expanded pop-up box will open. This is the box where we will craft our new template!

   A. At the top-left corner, you will see **Text message template**. Click the + button.

   B. On the right, click inside the **Template name** box to add a name for your new template.

   C. Add your text in the large box, using the # button to add merge fields if needed.

   D. Click the **calendar** icon to insert a booking link into your template.

   E. When you're done, click the **Insert Template** button to save your work.

4. You will be automatically sent back to the edit message page.

5. To use your new template, click the 🗐 icon again. Now, you can do the following:

    A. Scroll through the list to find your text template.

    B. Type the name of your text template.

6. Proceed with *Steps 3* to *13* of the *Text broadcasts* section.

How it works...

Overall, the use of templates in text broadcasting simplifies the message creation process, increases productivity, and strengthens the overall impact of your outreach efforts.

Broadcast reports

Reviewing your email broadcast reports is crucial for evaluating the effectiveness of your email campaigns and making informed decisions for future strategies. These reports offer valuable insights into key metrics such as open rates, click-through rates, and engagement levels.

By analyzing this data, you can identify which aspects of your email broadcasts resonated with your audience and which areas may need improvement.

Knowing your audience and understanding their behavior allows you to refine your content, optimize send times, and tailor your approach to better meet the preferences of your audience.

How to do it...

Follow these steps:

1. Click on the **COMMS** tab from the left-hand side navigation bar to open the menu. Choose **EMAIL BROADCASTS**.

2. Keap will display a list of previously sent and/or draft broadcasts. For this recipe, we will be looking for a **Sent** email.

3. Click the three black dots (ellipses) to the right of your email and choose **View Report**:

A blooming good bundle

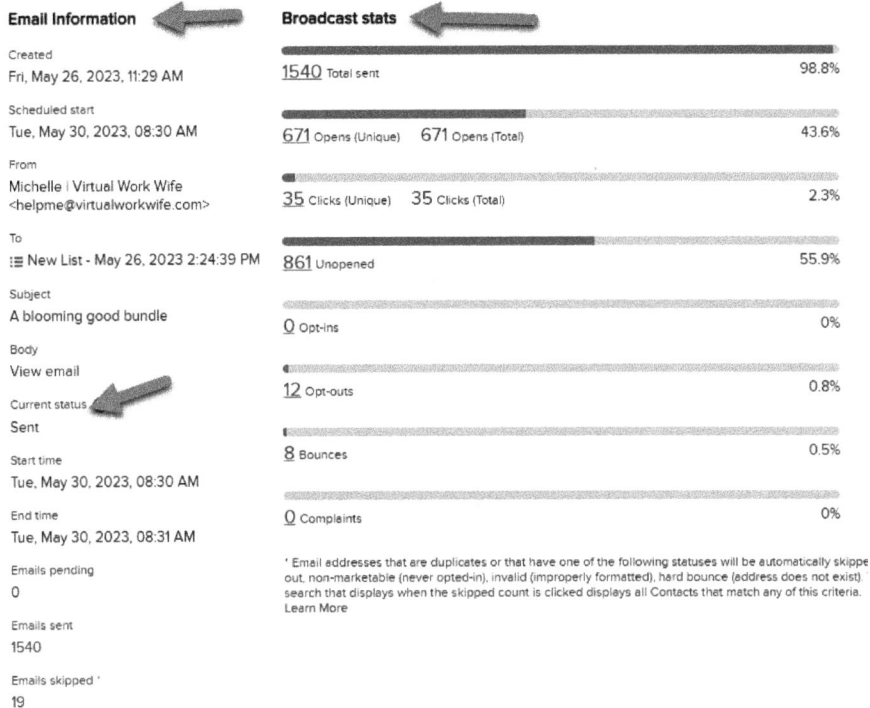

Email Information

Created
Fri, May 26, 2023, 11:29 AM

Scheduled start
Tue, May 30, 2023, 08:30 AM

From
Michelle | Virtual Work Wife
<helpme@virtualworkwife.com>

To
:≡ New List - May 26, 2023 2:24:39 PM

Subject
A blooming good bundle

Body
View email

Current status
Sent

Start time
Tue, May 30, 2023, 08:30 AM

End time
Tue, May 30, 2023, 08:31 AM

Emails pending
0

Emails sent
1540

Emails skipped *
19

Broadcast stats

1540 Total sent   98.8%

671 Opens (Unique)   671 Opens (Total)   43.6%

35 Clicks (Unique)   35 Clicks (Total)   2.3%

861 Unopened   55.9%

0 Opt-ins   0%

12 Opt-outs   0.8%

8 Bounces   0.5%

0 Complaints   0%

* Email addresses that are duplicates or that have one of the following statuses will be automatically skippe
out, non-marketable (never opted-in), invalid (improperly formatted), hard bounce (address does not exist).
search that displays when the skipped count is clicked displays all Contacts that match any of this criteria.
Learn More

Figure 4.16 – Email broadcast report

4. The email stats are separated into sections:

A. **Email Information**:

- **Created**: The date the email was created. This might be different from the date it was sent.

- **From**: Who the email was sent from.

- **To**: Who was included in the broadcast and how they were added (tags, groups, new search, and so on).

- **Subject**: The email subject line.

- **Body**: Clicking this link will display the email template.

- **Current Status**: This can be sent, scheduled, done, canceled, queued, and so on.

- **Start time**: The time the email started to be sent to recipients.

- **End time**: The time the email finished being sent.

B. **Current Status**: Each email status total that's displayed is linked so that you can review the list by clicking the link. It will display all contacts that match the criteria:

- **Emails pending**: How many emails were scheduled to be sent

- **Emails Sent**: How many got sent

- **Emails Skipped**: The number of emails that weren't sent

- **Emails Errored**: The number of emails that failed to send

> **Important note**
>
> Email addresses that are duplicates, have previously opted out, become non-marketable, reported spam, are improperly formatted, have hard bounced, or are empty are automatically skipped.

C. **Broadcast Stats**: These stats are located in the middle of the page and include both the number of contacts as well as the percentage of recipients for each stat. Clicking on the number of any stat will display the list of contacts for that stat:

- **Total Sent**: The total number of emails sent and percent

- **Open**: The overall open rate for the broadcast

- **Clicks**: Displays the number of contacts that clicked on a link

- **Unopened**: The percentage of emails not flagged as "open"

> **Note**
>
> Open rates are not exact and are only intended to give an estimate

- **Opt-ins**: Contacts that clicked on an email confirmation link

- **Opt-outs**: Contacts that unsubscribed from your list

- **Bounces**: Emails rejected by the receiving email server

- **Complaints**: The recipient marked the email as spam and reported it through their **email service provider (ESP)**, such as Google or Yahoo!

How it works...

Regularly reviewing email broadcast reports is an essential practice for enhancing the impact of your communication efforts, fostering continuous improvement, and ultimately achieving better results in reaching and engaging your target audience.

5
Managing Sales Pipeline

Managing follow-ups can become challenging as your business grows, resulting in overlooked tasks and missed opportunities. To maintain a professional image and seize every chance at converting leads into paying customers, every business can benefit from implementing sales pipeline automation. This ensures timely and consistent follow-ups and prevents a lack of communication, which as we all know leads to lost sales.

A sales pipeline is much more than just a visual representation of your sales process. It acts as the engine that propels your business forward. Each stage, from initial contact to closing the deal, presents an opportunity to create automation that saves you time and resources. Tracking and prioritizing leads, forecasting revenue, and identifying potential bottlenecks become effortless. This Keap feature isn't just a time saver; it's a game changer!

In this chapter, we'll explore the intricacies of setting up a sales pipeline automation system that effectively tracks lead progression from first contact to deal closing. We'll start by creating your sales pipeline, and then move on to determining key metrics to monitor, selecting triggers to propel leads through each stage, and setting up quotes and invoices to convert leads into paying clients.

This chapter contains the following recipes:

- Creating your sales pipeline
- Creating quotes
- Generating invoices
- Creating checkout forms

Creating your sales pipeline

Picture staying on task and on target, while never missing out on an opportunity again as you seamlessly move deals along your automated pipeline. Let's explore how this dynamic system can elevate your sales strategy to new heights!

Getting ready

Your pipeline automation will be highly customized to your business, based on your unique sales process. So, before you begin, you'll want to identify a few key elements for your sales process.

Think about all the steps you take a potential client through in your business. From the moment you meet them to the day they post about the work you did on social media, what are the steps of the journey? Here's an example of a basic sales journey:

1. New lead
2. Qualification
3. Initial contact
4. Proposal
5. Negotiation
6. Closing
7. Follow-up
8. Upsell
9. Referral

This is a good time to grab a whiteboard or a piece of paper and sketch out your client journey. This will be the blueprint we'll use to customize your pipeline in Keap.

Next, you'll need to identify what criteria or triggers you want to use to move leads from one stage to the next. Write this down or add it to your sketch so that you can reference it as you're building your pipeline.

Pipeline automation is powered by Keap's Easy Automations feature, which we will be covering in more detail in *Chapter 7*.

How to do it...

In this section, let's look at how to create your pipeline, add a deal to your pipeline, and move a deal to a different pipeline.

Creating your pipeline

1. Click on the **Sales** tab in the left-side navigation bar to open the menu and choose **Pipeline**.

> **Note**
> It is possible to maintain several pipelines simultaneously. Keap will display each pipeline in its own tab.

Figure 5.1 – Creating a new pipeline

2. Click on the three black dots (vertical ellipsis) to open the menu and select **Manage pipelines**.

3. A list of your existing pipelines will be displayed. Click the + **Add a new pipeline** link to continue.

4. There are three basic pipeline templates for you to choose from:

    A. **Sales pipeline**: Used for tracking and managing deals. This is a great choice if you're just getting started.

    B. **Project pipeline**: Used for tracking your progress on tools or products you want to develop.

    C. **Custom pipeline**: This is an excellent choice for businesses currently leveraging a pipeline with a robust set of established steps.

    For this exercise, we will be creating a sales pipeline.

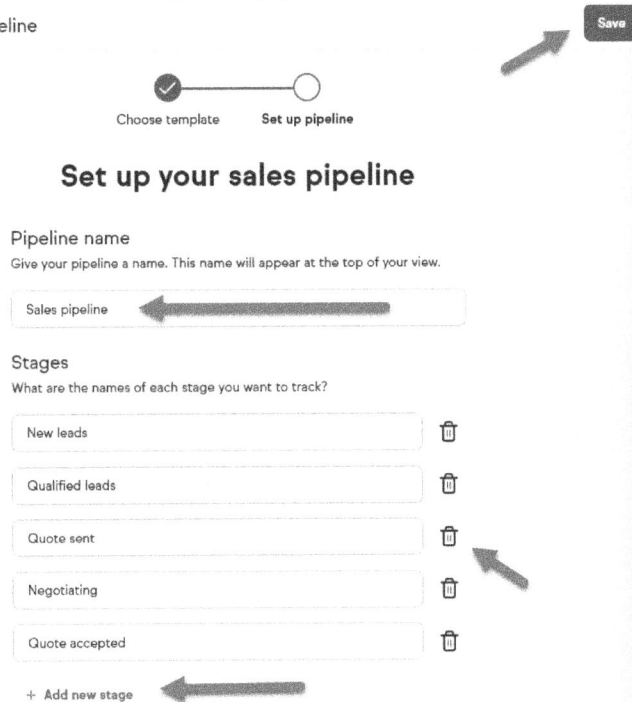

Figure 5.2 – Adding pipeline stages

5. The default sales stages have already been created for you. Edit the labels by clicking into the boxes and updating the names to fit your style.

6. You can delete a step by clicking the trash can symbol to the right of the field.

7. Click on + **Add new stage** below the fields to add new stage fields.

8. When you have added all the stages that you need for your unique pipeline, click the **Save** button in the top-right corner to return to the **Pipeline** main page. Your new pipeline will be added to the list.

Adding a deal to your pipeline

1. Navigate to your pipeline via the **Sales** menu.

2. At the stage where your deal should enter the pipeline, click + **Add a deal** to open the pop-up menu.

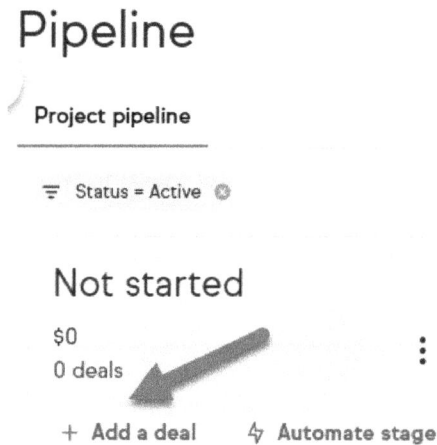

Pipeline

Project pipeline

⇗ Status = Active ⊗

Not started

$0
0 deals

⋮

+ Add a deal   ⚡ Automate stage

Figure 5.3 – Adding a deal

3. Confirm that the stage is correct or use the dropdown to change the stage.

4. Select all the contacts you want to create a deal for, or click the + **Add a new contact** option to create a new contact to be added.

5. Give your deal a name, set a value for **Deal value**, and add any additional notes for yourself or your sales team.

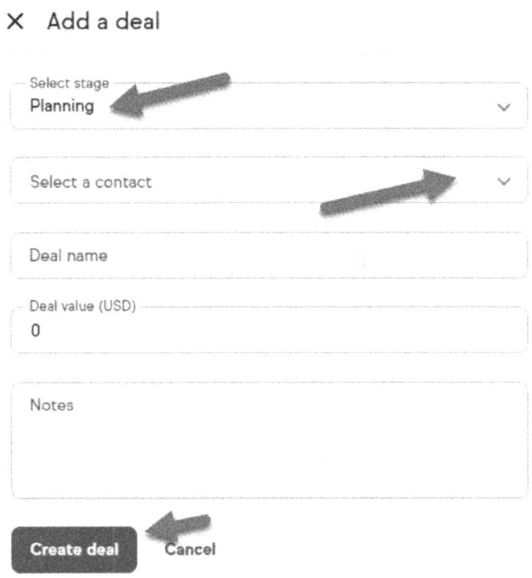

Figure 5.4 – Adding deal details

6. Click the **Create deal** button to save.

7. Once it has been saved, your new deal will pop up on the screen for you to review. You can now do things such as add notes, generate quotes or invoices, and send emails directly from your deal.

Moving a deal to a different pipeline

At times you may want to move deals between pipelines. This can be done by simply dragging the deal to the new pipeline, or by following these steps:

1. Click on the deal you want to move to open the popup.

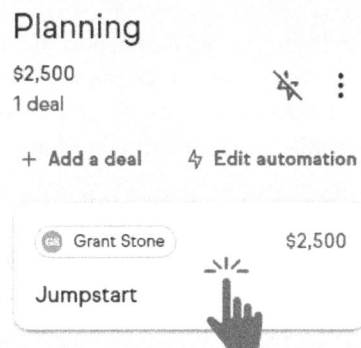

Figure 5.5 – Moving a deal

2. In the **Stage** dropdown, select the pipeline and stage you would like to transfer this deal to.

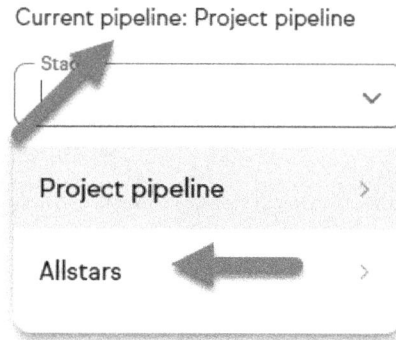

Figure 5.6 – Choosing a pipeline and stage

Your deal's activity is tracked and displayed at the bottom of the popup.

How it works...

Think of your pipeline as a visual representation of all your sales opportunities and where they are in your sales process at any given time. You'll be able to eliminate manual tracking processes such as checklists and spreadsheets because everything will be tracked—and easily reported on—in the digital system.

Individual deals can be moved through the sales pipeline, triggering automations that create tasks for your team or send emails and texts.

This not only eliminates the possibility of deals falling through the cracks but also allows you to forecast and project sales for the next week, month, or quarter.

Creating quotes

Keap's quote function gives you the ability to create a formal document or written statement for your potential buyers. It outlines the terms, conditions, and costs associated with a specific product or service that your buyer is interested in purchasing. A quote typically includes details such as prices, quantities, delivery timelines, and any other relevant terms and conditions.

Using quotes in your sales process is important for several reasons:

- **Clarity**: Quotes provide a clear and detailed overview of the products or services being offered, eliminating any confusion or misunderstandings that might occur.

- **Professionalism**: Providing a well-structured and professional quote shows your potential buyer your commitment to transparency and professionalism, instilling confidence in the buyer and enhancing your credibility.

- **Legal protection**: A quote serves as a legally binding document that outlines the terms of the sale. It can help protect both parties in case of disputes or misunderstandings by clearly stating the agreed-upon terms and conditions.

- **Negotiation**: Quotes are the basis for any negotiation. If your buyer has specific requirements or budget constraints, the quote can be a starting point for discussing any customizations or adjustments.

Using quotes in your sales process helps create a transparent, professional, and legally binding framework for the transaction. It facilitates effective communication, supports decision-making, and contributes to the overall success of the sales relationship.

Getting ready

You can create quotes and send them directly to your contacts. Your potential buyers can effortlessly approve your quotes by simply clicking the **Accept Quote** button. Plus, with a click of the **Convert to Invoice** button, your quotes can swiftly transform into payable invoices. Keep tabs on the status of all your quotes in your pipeline and receive notifications when they are accepted. Your pipeline is a one-stop shop for effortlessly managing your work. Access the quotes feature conveniently located under the **Money** tab alongside the invoicing function.

How to do it...

There are three ways to create a quote:

- From the **Sales** menu
- From a contact record
- From a deal that is in a pipeline, as outlined in the previous section

Creating a quote from the Sales menu

1. Click on the **Sales** tab in the left-side navigation bar to open the menu and choose **Quotes**.

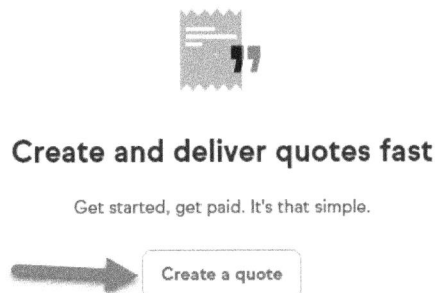

Create and deliver quotes fast

Get started, get paid. It's that simple.

Create a quote

Figure 5.7 – Creating quotes

2. If this is your first time, click on the **Create a quote** button to open the pop-up box. Otherwise, click the **Add a quote** button in the top-right corner.

3. In the **To** dropdown, you can do both of the following things:

    A. Search for the name of your potential buyer. Note: if you have multiple contacts with the same name you'll want to verify their info after selecting one to ensure it's the correct person.

    B. Click the + **Add a new contact** option to add a new lead.

Creating a quote from the contact record

1. Use the search feature to navigate to the desired contact.

2. Click **More**.

3. Then click **Add payment, invoice, or quote** and choose **Quote**.

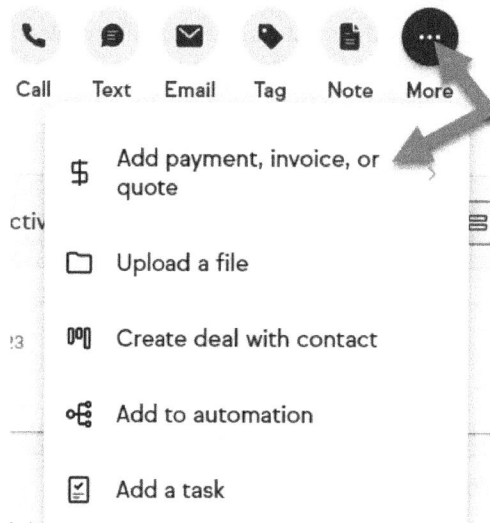

Figure 5.8 – Creating a quote from the contact record

Customizing quotes

The quotes feature allows you to easily add one or more products to your quote, set specific dates, and add any notes you want to share with your potential buyer:

1. Click the quote date to display the calendar and easily change your date.

Quote date

Mar 10, 2024

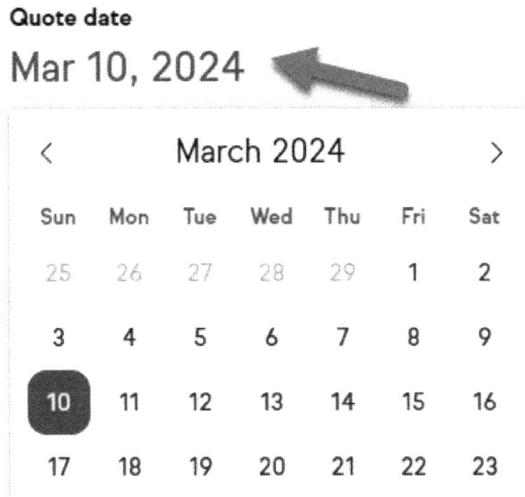

Figure 5.9 – Choosing a deal date

2. Click **+ Add a line item** to open the product search tool. Add an existing product or create a new one to add to your quote.

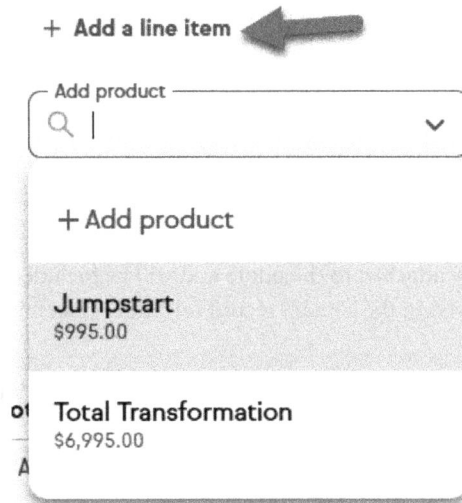

Figure 5.10 – Add existing or create new products

3. Once the product has been added to your quote, you can edit the price and quantity. This only affects the quote you are working on and does not change the product values anywhere else.

4. Use the **Notes** and **Terms** boxes to add any additional details to your quotes.

All quote activity is automatically saved and displayed at the bottom of the quote popup.

Deal activity

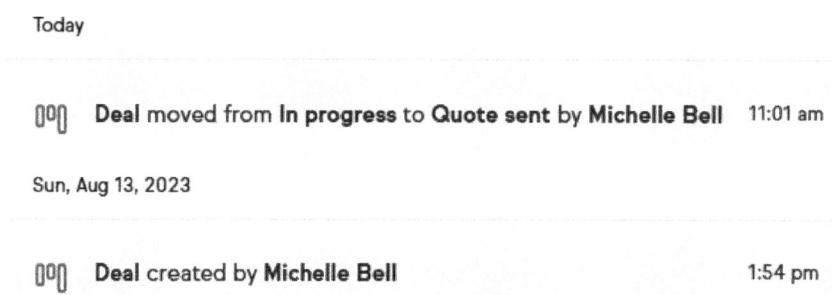

Today

〰️ **Deal** moved from **In progress** to **Quote sent** by **Michelle Bell**   11:01 am

Sun, Aug 13, 2023

〰️ **Deal** created by **Michelle Bell**   1:54 pm

Figure 5.11 – Deal activity

Attaching files to your quote

Attaching files to your quote can be a good idea if doing so enhances clarity, outlines customizations, or provides an extra level of professionalism. However, it's important to strike a balance and ensure that the attachments contribute meaningfully to your potential buyer's understanding without causing information overload.

1. Open your quote.

2. Click the **Attach a file** button.

3. Drag your attachment into the box or click the drag-and-drop link.

4. Your file is automatically attached to the quote and will be included in the email sent to your potential buyer, then saved in the contact record for future reference.

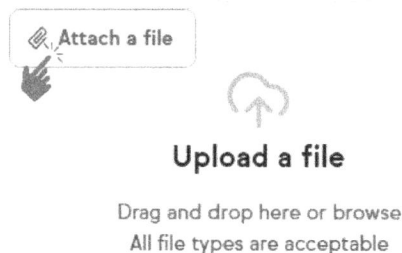

📎 Attach a file

Upload a file

Drag and drop here or browse
All file types are acceptable

Figure 5.12 – Attaching files to your quote

Sending your quote to the recipient

1. Click the **Next** button (in the top-right corner). This will open an email editor that is pre-populated with the recipient's information and the subject **your quote is ready to view**, along with some prewritten copy about the quote.

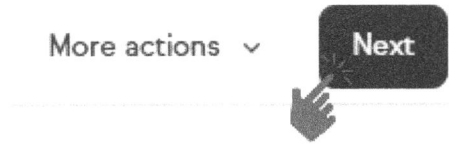

Figure 5.13 – Emailing your quote

2. You can now either customize the pre-written message to fit your personal style or leave it as is.

3. Toggle the **Signature** switch to display your signature details.

If you are now satisfied with your message, click **Send** to complete the email and send it to the recipient. If not, click anywhere on the screen outside of the email to exit the popup and return to your quote.

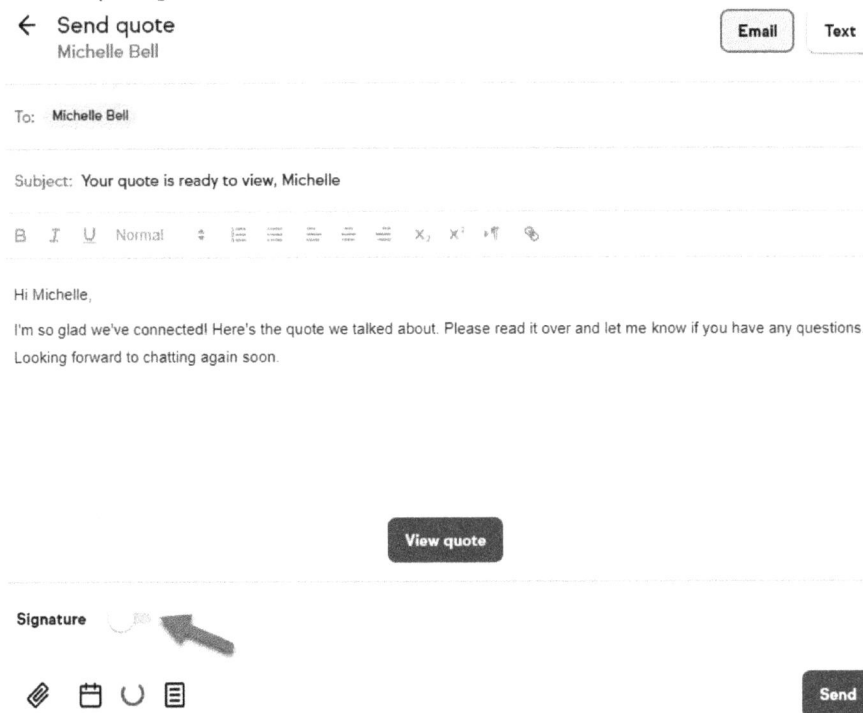

Figure 5.14 – Customizing your quote email

4. Clicking **X** to close your quote will take you back to the quotes page, where you will see your quote listed as having a **Draft** status.

5. To continue editing your quote, simply click on it to open the popup.

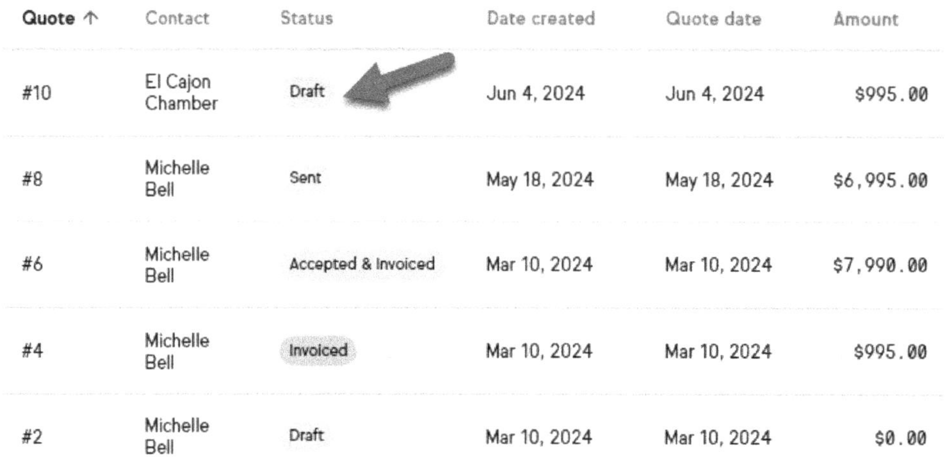

Quote ↑	Contact	Status	Date created	Quote date	Amount
#10	El Cajon Chamber	Draft	Jun 4, 2024	Jun 4, 2024	$995.00
#8	Michelle Bell	Sent	May 18, 2024	May 18, 2024	$6,995.00
#6	Michelle Bell	Accepted & Invoiced	Mar 10, 2024	Mar 10, 2024	$7,990.00
#4	Michelle Bell	Invoiced	Mar 10, 2024	Mar 10, 2024	$995.00
#2	Michelle Bell	Draft	Mar 10, 2024	Mar 10, 2024	$0.00

Figure 5.15 – Opening a quote from Draft

More actions

Within the popup for your quote, there is a **More actions** dropdown. This is where you will find additional options for managing your quote.

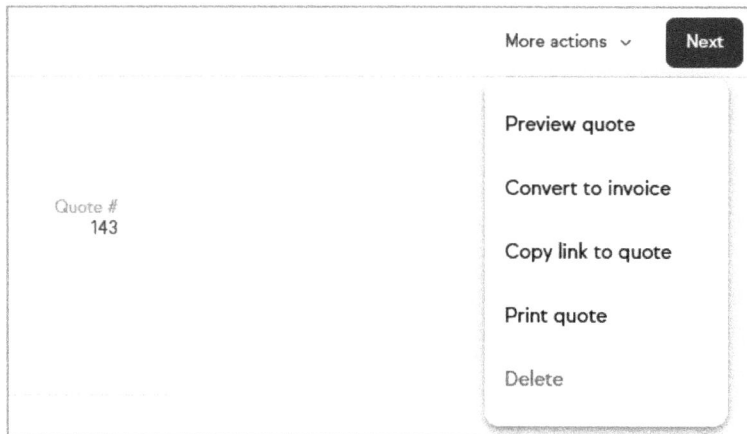

Figure 5.16 – Quote actions

- **Preview quote**: Displays what your client will see upon delivery

- **Convert to invoice**: Creates an invoice from your quote

- **Copy link to quote**: Copies the link to the quote so you can paste it to another location

- **Print quote**: Opens the printer options

- **Delete**: Removes quote record

How it works...

Quotes provide a crystal-clear overview of offered products or services, ensuring transparency and eliminating any potential confusion. The act of presenting a well-structured and professional quote showcases your commitment to transparency and professionalism, instilling confidence in potential buyers and enhancing your overall credibility.

Generating invoices

Timely and accurate invoicing contributes to positive customer relationships. Keap provides businesses with the tools to efficiently manage invoicing, reducing the likelihood of errors and improving overall customer satisfaction.

Getting ready

Here's how invoices are typically used within Keap:

- **Record-keeping and documentation**: Invoices serve as a record of transactions between the business and its customers. They provide a documented history of purchases, payments, and any related financial transactions.

- **Order processing**: Invoices are often linked to sales orders and quotes. When a customer accepts a quote or places an order, Keap can generate an invoice automatically, streamlining the order-to-invoice process. This ensures accuracy and efficiency in managing sales transactions.

- **Payment tracking**: Monitoring outstanding balances is a critical step in running a business. This helps in managing cash flow, identifying overdue payments, and sending reminders to customers for timely settlements.

Within the Keap invoicing system, you can send professional invoices and track their statuses easily. Easy automations allow you to send a follow-up email once the customer pays an invoice and move the client into a fulfillment workflow.

How to do it...

When you create and modify your invoice, any adjustments you make will be saved automatically.

Creating an invoice

1. Click on the **Sales** tab in the left-side navigation bar to open the menu, and then choose **Invoices**.

2. Click the **Add** button in the top-right corner, then choose **Invoice**.

3. Use the **Billed to** dropdown to search for and select an existing contact or click the + **Add a new contact** link to add a new record.

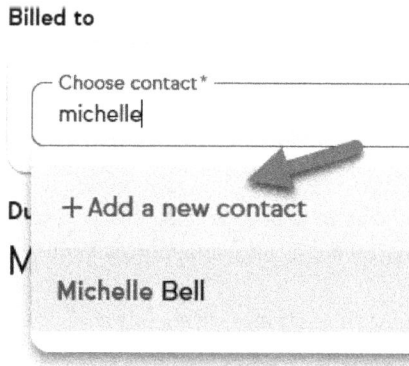

Billed to

Choose contact *
michelle

+ Add a new contact

Michelle Bell

Figure 5.17 – Adding a contact to an invoice

4. Click on the date to open the calendar feature and change the **Due by** date.

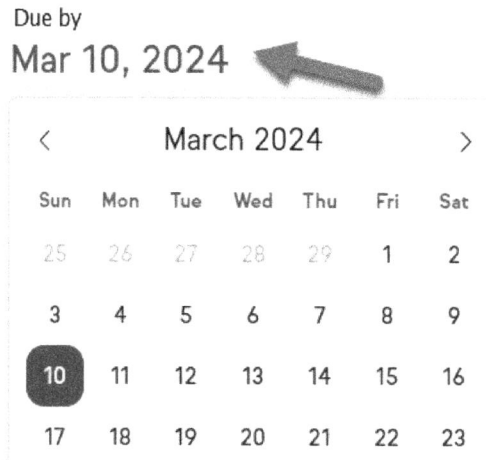

Due by
Mar 10, 2024

<		March 2024				>
Sun	Mon	Tue	Wed	Thu	Fri	Sat
25	26	27	28	29	1	2
3	4	5	6	7	8	9
10	11	12	13	14	15	16
17	18	19	20	21	22	23

Figure 5.18 – Changing the Due by date

5. Click + **Add a line item** to add products to your invoice.

6. Product prices will automatically load to your invoice. You can edit the price and/or quantity of a product by clicking the edit icon for the line item.

Figure 5.19 – Editing the price and quantity

Adding sales tax

1. After adding a product to an invoice, you can add sales tax, if applicable, by clicking + **Add sales tax**.

2. Select the appropriate state from the dropdown or click + **Create a new sales tax**.

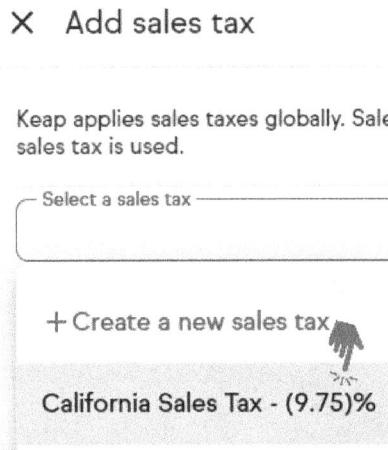

Figure 5.20 – Adding sales tax

> **Note**
> Each state has its own rules for when or if you need to collect state taxes. For more information regarding your tax obligation, contact your CPA or visit your state's website for the most up-to-date answers.

3. Click **Save**.

The tax rate will be applied to the specific line item(s) and any taxes included on the invoice will be displayed in the summary at the bottom of the invoice.

Applying discounts and requesting deposits

You can add a discount or request a deposit from buyers when sending invoices:

1. Click on **Add a discount** or **Request deposit**.

2. Add a discount or deposit value:

    A. Set a predefined dollar amount.

    B. Set a percentage of the invoice total.

Subtotal	$7,994.00
Add a discount	
Amount paid	$0.00
Total due	**$7,994.00**
Request deposit	

Figure 5.21 – Setting discounts or requesting deposits

3. If needed, you can remove the discount or deposit by clicking the trash can icon.

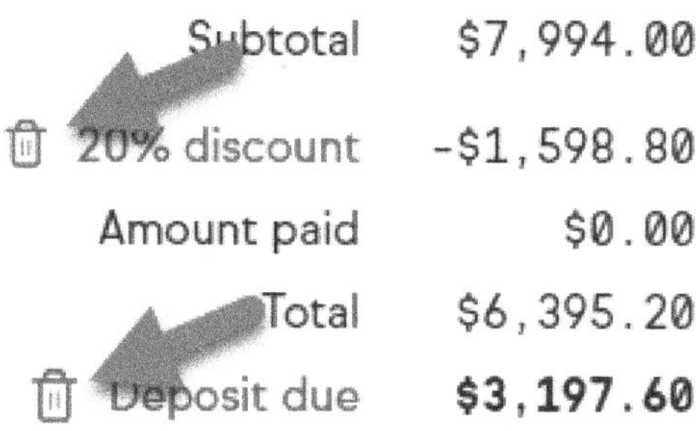

Subtotal	$7,994.00
🗑 20% discount	-$1,598.80
Amount paid	$0.00
Total	$6,395.20
🗑 Deposit due	**$3,197.60**

Figure 5.22 – Removing discounts or deposits

Accepting online payments

Before you can accept payments, you will need to set up your payment processor. This step was covered in *Chapter 2*.

1. Toggle **Accept credit cards** or **Accept PayPal** to enable or disable each payment method.

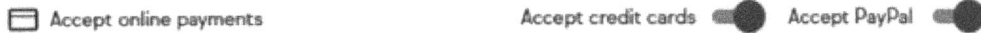

Figure 5.23 – Turning payment methods on or off

A. If both options are enabled, your customers will be able to choose the payment method they prefer.

> **Note**
> If you have multiple payment processors set up in Keap, the invoice will charge the credit card using the processor you have set as your default.

B. If no payment options are enabled, your customer will only be able to print an invoice.

Adding notes and terms

You have the option to input notes and/or terms at the bottom of your invoice by adding them to the provided boxes.

Figure 5.24 – Adding notes and terms to your invoice

In the context of invoicing, terms refer to the agreed-upon conditions and stipulations regarding payment between you, the seller, and your potential buyer (recipient of goods or services).

These terms outline crucial details such as the payment due date, any applicable discounts for early payment or penalties for late payment, and the method of payment accepted. Common terms include **net 30**, indicating that payment is due within 30 days.

Clear and well-defined terms, which you can conveniently input at the bottom of your invoice, help establish expectations and ensure a smooth and transparent financial transaction between the parties involved.

Reviewing and sending your invoice

1. Click **Next** when you're finished creating your invoice.

2. You will now see several options for putting the invoice into the hands of your potential buyer:

   A. **Email it**: Opens the email editor

   B. **Text it**: Sends a text using your Keap business line

   C. **Share a link**: Copies the invoice link so you can use it in another location

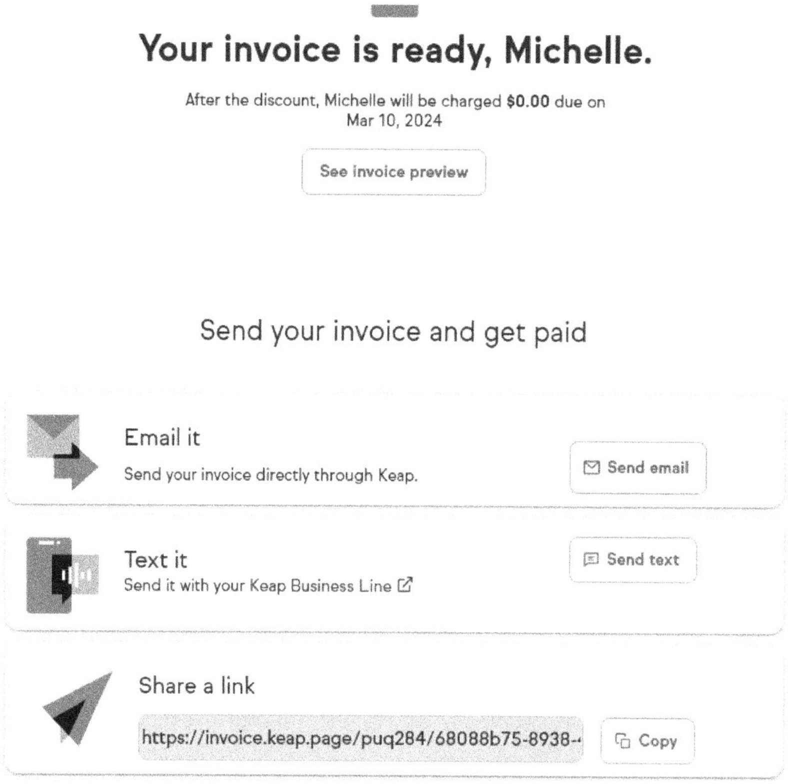

Your invoice is ready, Michelle.

After the discount, Michelle will be charged **$0.00** due on
Mar 10, 2024

See invoice preview

Send your invoice and get paid

Email it
Send your invoice directly through Keap.

☐ Send email

Text it
Send it with your Keap Business Line ⤤

☐ Send text

Share a link
https://invoice.keap.page/puq284/68088b75-8938-

☐ Copy

Figure 5.25 – Options for sending your invoice

More actions

Within the **More actions** dropdown, you'll find options such as **Add a payment**, **Preview invoice**, **Copy invoice link**, **Print invoice**, and **Delete**.

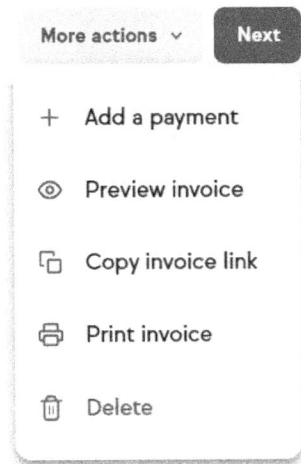

Figure 5.26 – Invoice actions

- **Add a payment**: Allows you to manually apply payments to the invoice
- **Preview invoice**: Displays what your client will see when they view the invoice
- **Copy invoice link**: Copies the link to the invoice so you can paste it to another location
- **Print invoice**: Lets you print a copy or save it as a PDF

How it works...

Invoices play a pivotal role in your business. Keap's invoicing system ensures a streamlined and effective approach to managing financial transactions, promoting clarity, and facilitating smoother communication between businesses and their clients.

Keap's invoicing system empowers users to send professional invoices effortlessly, with easy automations enabling follow-up emails upon payment and seamless integration into fulfillment workflows.

Creating checkout forms

The **Checkout Forms** feature of Keap Max and Pro enables you to collect payments from your customers without the need for an invoice. Simply create your customized form, then share it with your customers, post it to social media, or link it on your website.

Getting ready

Before you can use checkout forms, you will need to set up your payment processor. This step was covered in *Chapter 2*. You can click on the **Setup payment processing** button as shown here:

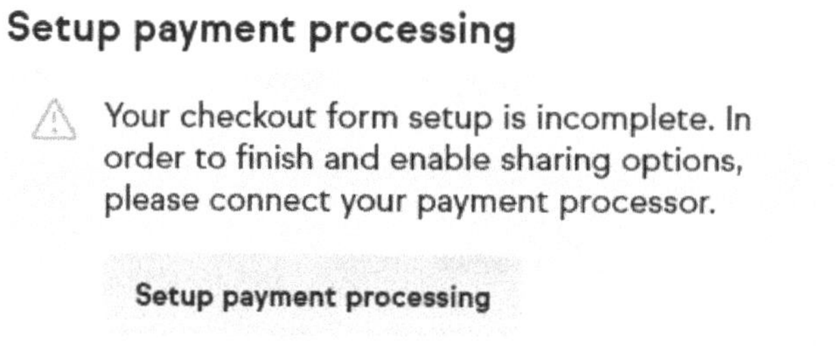

Setup payment processing

⚠ Your checkout form setup is incomplete. In order to finish and enable sharing options, please connect your payment processor.

Setup payment processing

Figure 5.27 – Setup payment processing warning

How to do it...

In this section, we'll cover how to create a checkout form, add an upsell, add a promo code, and add a thank you page.

Creating a checkout form

1. Click on the **Sales** tab in the left-side navigation bar to open the menu, and then choose **Checkout Forms**.

2. Click the **Add a checkout form** button in the top-right corner to start your form.

3. Click the **+ Add product** button to open the product search tool. Add an existing product or create a new one to add to your checkout form.

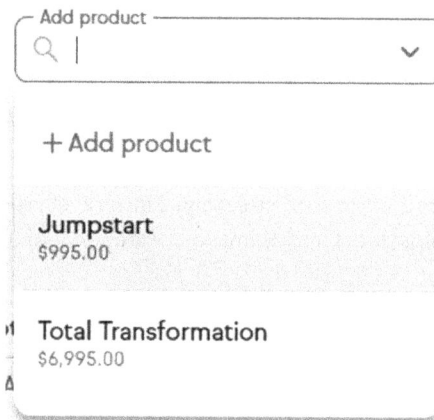

┌─ Add product ──────────────────
│ 🔍 | ⌄
│
│   + Add product
│
│   Jumpstart
│ $995.00
│
│   Total Transformation
│ $6,995.00

Figure 5.28 – Add existing or create new products

Adding an upsell

You can display up to two additional products on your order form as upsells. They will display under the **You may also like** header on your form. These are optional products that the potential buyer can add to their order at the time of sale.

To add an upsell product, use the following method:

1. Click + **Add upsell** to open the product search tool. Add an existing product or create a new one to add to your checkout form.

2. Repeat *step 1* to add a second upsell to your form.

Upsell

Would you like to add an upsell product to the checkout form?

payment $219.90

                                                      + Add sales tax

Figure 5.29 – Adding upsells to your form

It's common practice to discount the price of products used as an upsell. To change the product price, use the following steps:

1. Click the edit icon for the upsell. You can now do the following:

   A. Change the name of the product.

   B. Add a description to create excitement about the upsell.

   C. Change the price.

   D. Decide whether you want a **One-time** or **Recurring** charge. If you select **Recurring** there will be additional criteria to choose, such as frequency of charge and payment schedule.

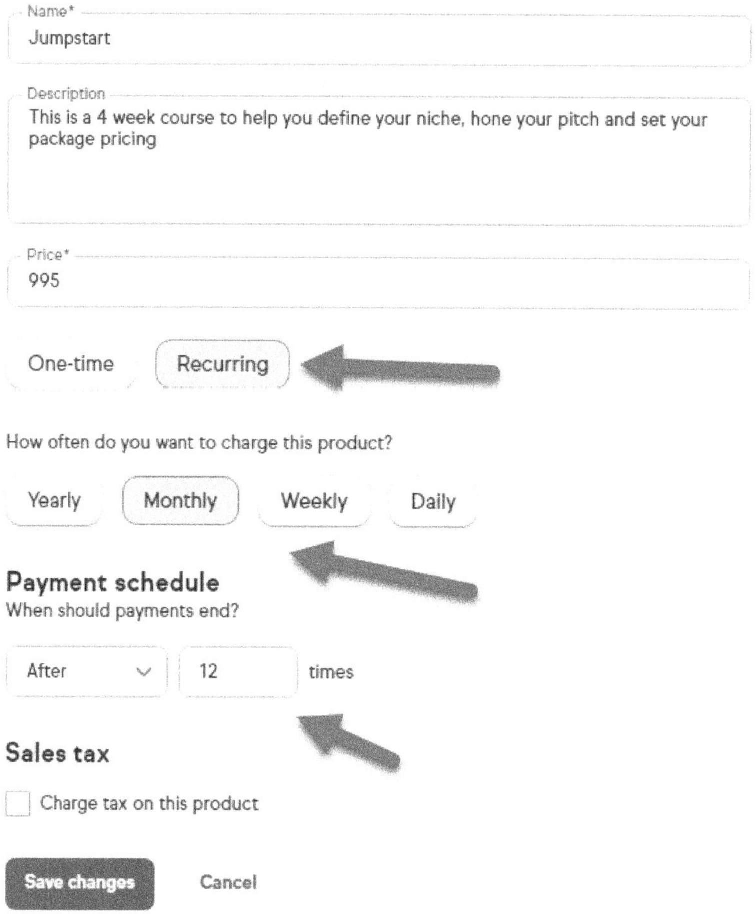

Figure 5.30 – Setting recurring charge criteria

    E.   Charge tax for the item.

2.   When done, click **Save changes** to return to your checkout form.

Adding a promo code

Give your potential buyers extra incentive to make a purchase by offering a discount promo code for use on Keap's checkout forms. Promo codes are a powerful tool you can use to generate exclusive discounts tailored for specific leads or contacts who have reached a particular stage in your pipeline.

Beyond the enticing discounts, promo codes offer you a huge advantage when it comes to comprehensive reporting. Track the effectiveness of various marketing initiatives, monitor referrals, or evaluate the performance of affiliates by assigning specific promo codes for their use. This feature provides valuable insights into which strategies yield the most business, enhancing your overall marketing intelligence.

To add a promo code, use the following steps:

1. Click + **Add promo code** to open the popup.

2. Create your code name (using all caps helps distinguish it from other text).

3. Choose a type and set the value as one of the following:

   A. A percentage of the total order

   B. A set dollar amount

4. Click **Save promo code** to close the popup.

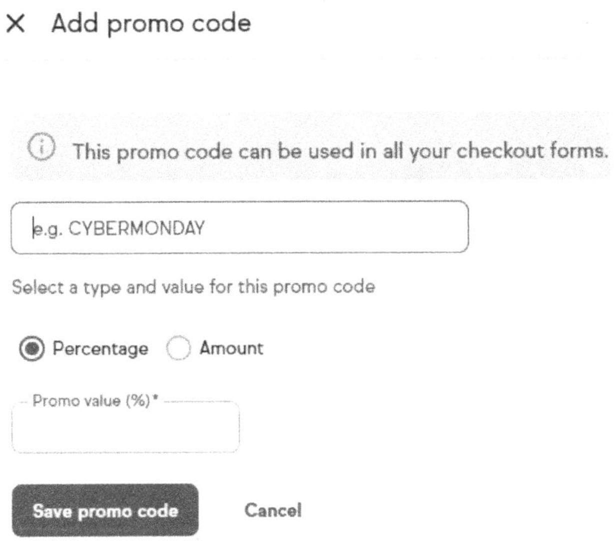

X   Add promo code

ⓘ This promo code can be used in all your checkout forms.

e.g. CYBERMONDAY

Select a type and value for this promo code

◉ Percentage   ◯ Amount

Promo value (%) *

Save promo code   Cancel

Figure 5.31 – Promo code setup

5. If you have not set up a payment processor, you will be prompted to do so. Refer to *Chapter 2* for instructions.

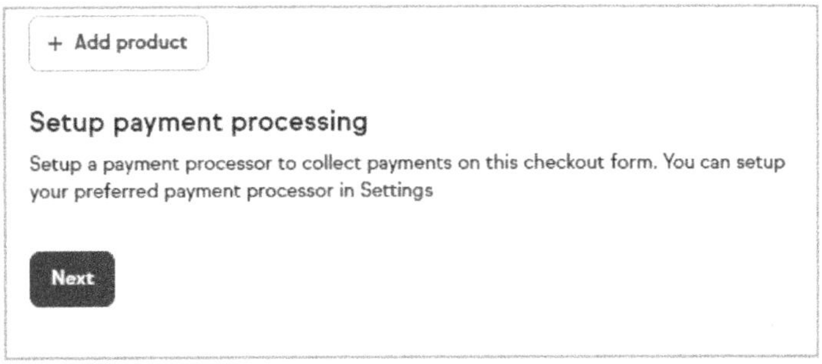

+ Add product

Setup payment processing

Setup a payment processor to collect payments on this checkout form. You can setup your preferred payment processor in Settings

Next

Figure 5.32 – Payment processing warning

6. Click the **Next** button to continue.

7. Create an internal name for your form.

8. Add a headline that will display at the top of your form.

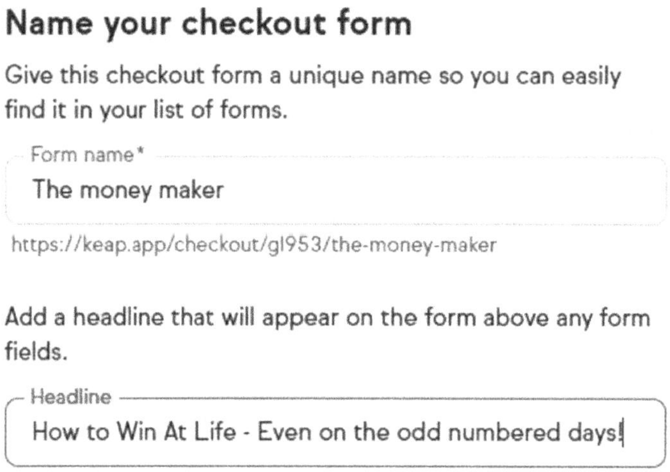

Name your checkout form

Give this checkout form a unique name so you can easily find it in your list of forms.

Form name*

The money maker

https://keap.app/checkout/gl953/the-money-maker

Add a headline that will appear on the form above any form fields.

Headline

How to Win At Life - Even on the odd numbered days!

Figure 5.33 – Checkout form name and headline

9. Click the **Next** button to continue.

10. Customize the appearance of your checkout form by opting to showcase your company logo and choosing page background and button colors that align seamlessly with your brand.

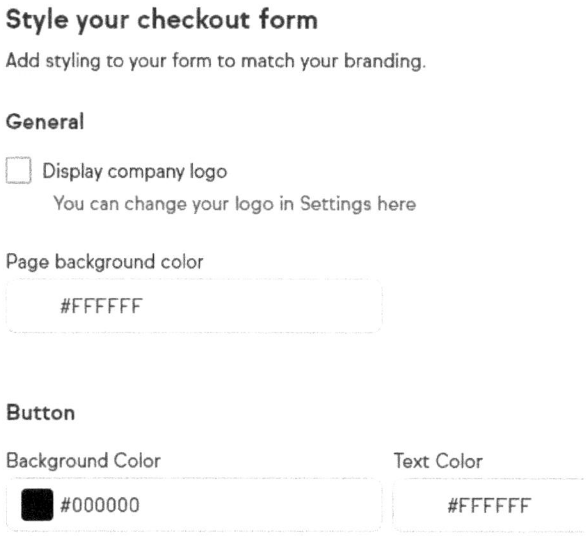

Style your checkout form

Add styling to your form to match your branding.

General

☐ Display company logo
You can change your logo in Settings here

Page background color

#FFFFFF

Button

Background Color

■ #000000

Text Color

#FFFFFF

Figure 5.34 – Checkout form style

Automating follow-ups for your checkout form

Following up with your buyers is easy with checkout automation. There are four options for automating your checkout follow-ups:

- **Send email to contact with product**: Let the customer know that their new product is on the way and detail what happens next.

- **Create a follow up task for yourself**: Alert your team (or yourself) that a new purchase has been made and start your fulfillment process.

- **Tag contacts with Snazzy Checkout Form**: Add buyers to a unique tag to segment them from the rest of your list so you can send targeted follow-ups later. This can be compared to requesting a testimonial or sending a special coupon code for repeat orders.

- **+ Create custom automation**: Create a more complex follow-up, such as drip emails that stop after an action has been taken, for example.

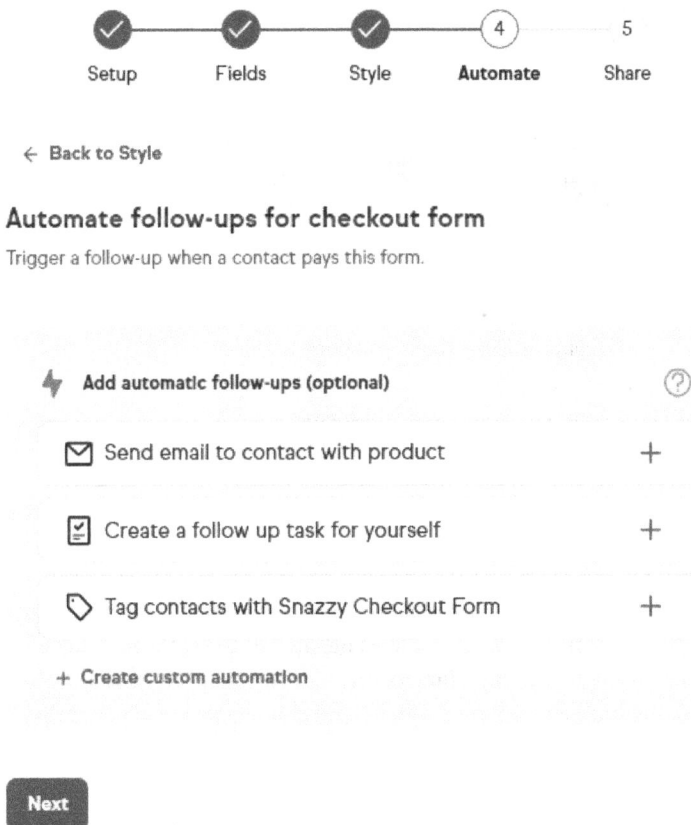

Figure 5.35 – Automate checkout follow-ups

Ideally, you will deploy one or more of these options to maximize your automations and guarantee that your new client has an awesome buying experience.

Adding a thank you page

Customize the appearance of your checkout form by opting to showcase your company logo and choosing page background and button colors that align seamlessly with your brand:

1. Click the drop-down menu to choose a page:

   A. **Default thank you page**: This is a basic page provided by Keap. It has very little styling but is quick and easy to deploy.

   B. **Redirect to another page (URL)**: This option requires you to create your own thank you page, typically on your website.

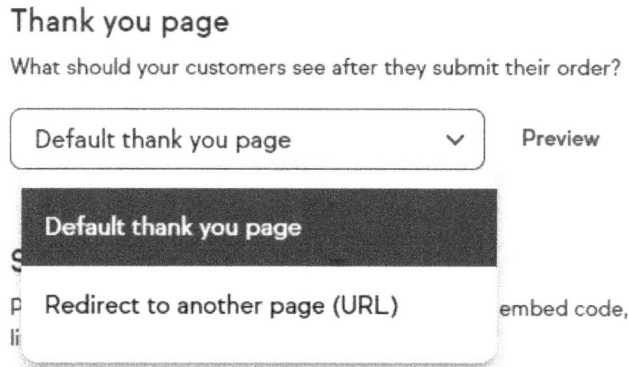

Figure 5.36 – Thank you page options

2. Click **Save** and **Exit** to complete your checkout form.

Your checkout form is now live and can be shared using a link or by embedding it on your website!

How it works...

The Checkout Forms feature offers significant value by simplifying the collection of payments without the need for invoices. This streamlined process allows businesses to create customized forms, share them effortlessly, and engage with potential buyers through various channels. The checkout form's versatility extends to adding products, optional upsells, and enticing promo codes, providing an extra layer of appeal to potential buyers.

The ability to customize the form's appearance ensures brand alignment, creating a professional and cohesive user experience. Overall, the Checkout Forms feature enhances efficiency, increases flexibility, and contributes to a more streamlined and customer-friendly sales process.

Part 3: Sales Pipeline Management

Here, you'll learn how to effectively manage your sales pipeline, track leads and opportunities, and optimize your sales process using Keap's built-in tools.

- *Chapter 6, Marketing Forms and Landing Pages*

6
Marketing Forms and Landing Pages

It's time to discover the distinctions between landing pages and forms, as well as the optimal scenarios for utilizing them within Keap's framework.

Forms serve as online data collection tools, commonly found embedded within web pages or presented as popups. They facilitate the gathering of crucial information, such as names, page addresses, and phone numbers, serving various purposes, including lead capture, sign-ups, and feedback solicitation.

Conversely, landing pages are standalone web pages strategically designed to achieve specific objectives, such as prompting visitors to make purchases, subscribe to newsletters, or download resources. They are meticulously crafted to captivate attention and drive targeted actions, without all the distractions of a traditional webpage.

Effective landing pages feature compelling headlines, engaging content, and persuasive calls to action, and often incorporate testimonials and social proof.

While forms are instrumental in data collection, landing pages are tailored to convert visitors into leads or customers. Forms can be integrated across multiple pages of a website, whereas landing pages stand alone, focused, and optimized for driving desired conversions.

In this chapter we'll cover the following topics:

- Building, styling, and publishing public forms
- Building, publishing, and using internal forms
- Creating landing pages

Technical requirements

For this chapter, the following are helpful, but not required:

- The hex codes for your preferred colors

- A preferred font to use consistently for your headers

- A preferred font to use consistently for your text

- Your logo or branding materials that will be used

Public forms

Public forms are typically tailored for external use, mainly targeting customers or potential leads. These forms should be easily accessible to anyone browsing your website, your landing pages, or even your social media profiles.

They serve the purpose of collecting vital information from visitors, such as contact details, inquiries, or requests for more information. Examples of public forms can be found on your **Contact us** page, lead capture forms, newsletter sign-ups, and request-a-quote forms. The data collected via these public forms is seamlessly integrated into Keap, enabling your sales and teams to effectively follow up with leads or customers.

Getting Ready

To create web forms effectively, it is essential to gather the following prerequisites to build a visually appealing and thoughtful funnel that gets people to your goal:

- **Objectives**: Clearly outline the purpose of your web form. Determine what information you want to collect from your audience and how you plan to use it.

- **Required Fields**: Decide which fields are essential for your web form. Common fields include name, page address, phone number, and any specific data relevant to your goals.

- **Compelling Copy:** It is important to clearly communicate the value of what you are offering and explain how the information will be used. Use persuasive language to encourage participation and instill trust in your audience.

- **User-Friendly Layout**: Consider things such as placement of form fields, readability of text, and visual appeal. Aim for a clean and intuitive design that guides users through the form completion process effortlessly.

Whether it's lead generation, event registrations, or feedback gathering, having a clear objective will guide the design and implementation of your web form. By preparing these key elements beforehand, you will be well equipped to create effective web forms quickly and start capturing that valuable information and driving engagement with your audience.

How to do it ...

1. Click on the **MARKETING** tab in the left-side navigation bar to open the menu and choose **FORMS**.

2. From here, you can choose to **Create form** or connect to external tools such as Typeform or WordPress. For this recipe, we will be clicking on the **Create form** button (top right corner).

3. You have three options: create an internal form, create a public form, or choose to start from a template. We're going to select **Public form**.

4. Click the **Start building your form** button.

> **Note**
>
> Forms are built in four sections; Build, Style, Automate, and Publish. Keap will auto-save your progress as you go.

5. Set your **Form name**.

6. Add a **Headline**. This is an optional field that will be displayed at the top of your form.

7. Customize the **Button text**. The default text is **Submit**, but it is highly recommended that you set your own value so that it aligns with your copy and objective.

The fields included in public forms vary depending on the template chosen. Typically, they include first name, last name, email address, and phone number. Additional fields can be added or default fields removed to tailor the form so that it precisely gathers the information you desire.

Building public forms

In Keap, building public forms is a streamlined process, empowering you to effortlessly create and customize forms and start capturing all those juicy leads!

Adding fields

1. Click the **Add fields** button below the list of already added fields.

Fields

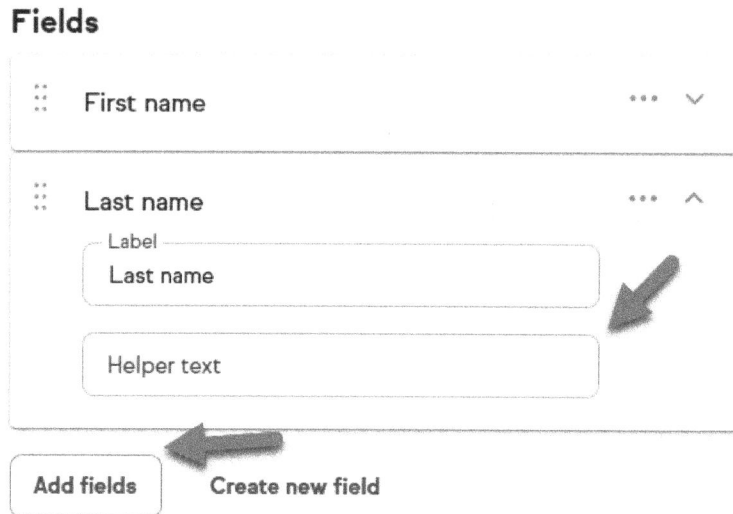

Figure 6.1 – Adding fields

2. Select the desired field by scrolling the list or typing in the search box.

3. Customizing your fields will ensure the user understands what you require:

    I.   Click on the dropdown arrow to the right of the field name.

    II.   Change the label to suit your needs.

    III. Adding Helper text offers additional information, instructions, or examples to help users understand what is expected in each field.

4. Repeat step 2 until you have added all your desired fields.

5. If you are unable to find a field that fits your needs among the standard fields, you have the option to **Create new field** by clicking the provided link:

    I.   Name your field.

    II.   Choose a field type from the dropdown.

    III. Click **Create Field** to save your work.

    IV.   Your new field will be automatically added to your form.

Editing and removing fields

To remove or change a field on your form, follow these steps:

1. Click the ellipsis to the right of the field name.

Figure 6.2 – Deleting a field

2. Use the toggle to make the field required or not required.

3. Set your helper text to always visible or not using the toggle.

To delete fields, follow these steps:

1. Click the ellipsis to open the menu.

2. Click the **Delete** button.

To move form fields, follow these steps:

1. Click and hold the field you want to move.

2. Drag the field up or down and then release the mouse button to place it above or below other fields.

Styling public forms

You can customize the appearance of your form by selecting options such as displaying your logo, configuring the page background color, adjusting the button background color, modifying the button text color, and aligning the buttons to match your personal branding guidelines.

To begin styling your form, follow these steps:

1. Click the **Next** button to move to **Style**.

2. Choose whether or not to display your logo by checking your unchecking the box.

3. Set your page background color in one of the following ways:

    A. Type your specific color code in the box.

    B. Click on the color patch to open the color mixer tool. Use your mouse to click on the area of color you want to select.

Page background color

#FE019A

Figure 6.3 – Using the color mixer

4. Set your button's background and text color using the same method as step 2.

5. Choose how you want to align your button on the form:

    A. Left – The button will appear in the lower-left corner of your form

    B. Right – The button will appear in the lower-right corner of your form

    C. Center – The button will appear at the bottom of your form

    D. Full – This will stretch your button as wide as your form

6. Click **Next** to continue to **Automation**.

Automating public forms

You can create automations that are triggered whenever someone submits your form. This can greatly improve your lead conversion and speed up your sales cycle.

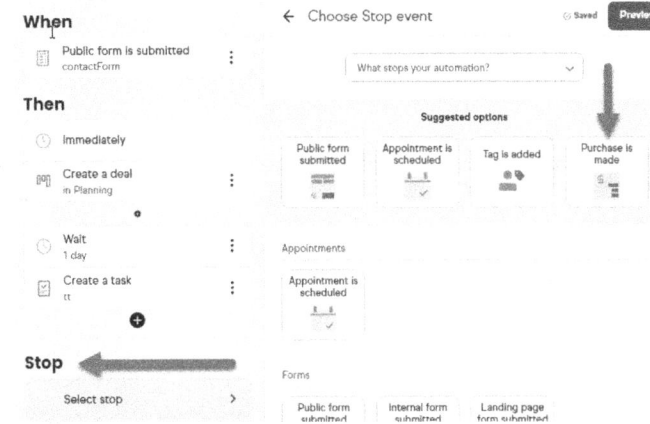

Figure 6.4 – Adding and editing an automated follow-up

For this recipe, let's select **Send contact 48-hour follow up email**:

1. Click the + next to **Send contact 48-hour follow-up email**.

2. The automation builder will slide open from the right with a pre-named automation and the **When** and **Then** values already set.

3. You can change the wait time by clicking the **Wait** field to open the editor and adjusting the duration. You can choose from days, weeks, months, minutes, or hours:

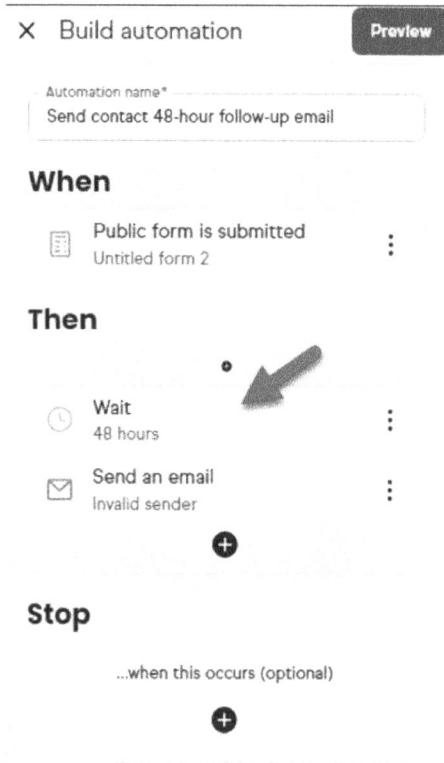

Figure 6.5 – Using the Wait field

4. When you are satisfied with your timing, click the **Next** button to return to the editor.

5. Next, click on **Send an email** to review and/or make changes:

    I.   Click the **From** section to change the sender.

    II.   Click the **Subject** to begin typing in the field.

    III.   Use the editing tools to change the body of the email.

    IV.   Use the **Close** button to return to the form builder.

Publishing a public form

Once your form has been styled and you have added automation, you are ready to move on to publishing. To do this, follow these steps:

1. Click the **Next** button (in the lower left corner) to continue.

2. Choose a **Thank you page** option from the drop-down menu and use the default page provided by Keap. Then, choose **Redirect to another page (URL)** if you want to send your viewer to a custom page or website. Type the complete URL in the provided box. Then you need to select your sharing options.

3. You have several options for sharing your newly minted form:

    A. **Share a link**: Good for using on social media, text messages, and adding to pages.

    B. **Share on Facebook**: Click on the share icon and log in to your Facebook account. You can add a comment to your message on the pop-up menu.

    C. **Share on Twitter**: Click on the share icon and log in to your Twitter account. You can add a comment to your message on the pop-up menu.

    D. **Add it to your website**: Using the provided code, you can embed your form directly on your website.

> **Note**
>
> Before you save and exit your form, use the **Try it out** button to test your form. You will get a sense of the look and feel of the form and confirm that your test subject was effectively added to any automation.

4. Click the **Save and exit** button to finalize and publish your form.

Deleting public forms

Deleting public forms is a straightforward process that involves removing unwanted or outdated forms from your system. By deleting unused forms, you can declutter your form library and ensure that only relevant forms are available for use. Additionally, deleting obsolete forms helps maintain data integrity and prevents users from accessing outdated or irrelevant information.

Before deleting a public form, it is important to review its usage and ensure that any associated data or submissions are appropriately archived or transferred to prevent data loss. Once confirmed, deleting a public form can be done quickly and easily within your form management settings.

To delete the form, follow these steps:

1. Click on the **MARKETING** tab in the left-side navigation bar to open the menu and choose **FORMS**.

2. Navigate to the form you want to delete and click the ellipsis on the far right.

3. Click **Delete** form.

4. You will be asked to confirm that you want to delete the form.

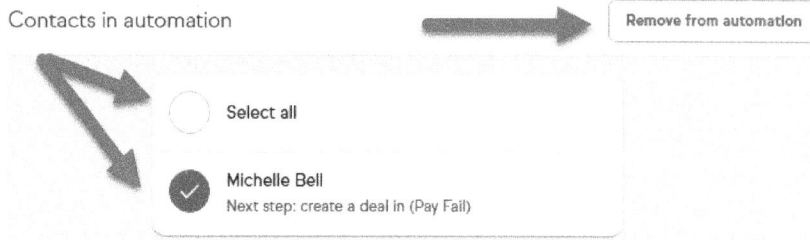

Figure 6.6 – Deleting a public form

5. Click **Delete** to confirm.

How it works...

Overall, public forms streamline the process of gathering information from external sources and facilitate efficient data management within the CRM system. They are your go-to tool for collecting information and getting leads into your sales funnel. By collecting data such as first name, last name, page address, and phone number you can create an automated journey that brings them one step closer to becoming loyal customers.

Deleting public forms once they are no longer useful is a straightforward process to declutter your form library and maintain data integrity.

Internal forms

Internal forms are a quick and easy way to get data into your CRM when a public form isn't needed. They provide a structured template for gathering information from internal users, help standardize procedures to improve data quality through validation checks, and facilitate collaboration and workflow automation within the organization, ultimately enhancing efficiency and productivity.

How to do it...

1. Click on the **MARKETING** tab in the left-side navigation bar to open the menu and choose **FORMS**.
2. Click the **Create form** button (in the top right corner).
3. Select the **Internal** option, then select **Start building your form**.

Building internal forms

Unlike the public form builder, the internal form builder only has two sections: Build and Automate.

Adding fields

In Keap, adding fields to a form is intuitive and efficient, allowing you to easily customize your forms so you capture the right information for the task at hand:

1. Click the **Add fields** button below the list of already added fields.

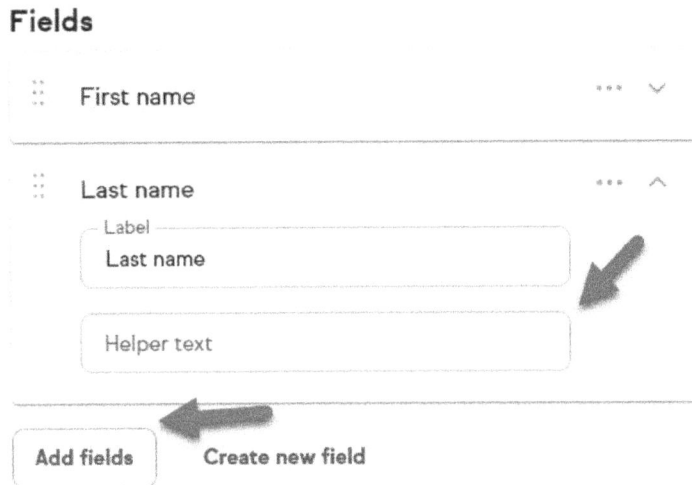

Figure 6.7 – Adding fields to an internal form

2. Select the desired field by scrolling the list or typing in the search box.

3. Customize your selected field:

    I.   Click on the drop-down arrow to the right of the field name.

    II. Change the label to suit your needs.

    III. Adding helper text provides additional information, instructions, or examples to help users understand what is expected in each field.

4. Repeat *step 2* until you have added all your desired fields.

5. If you are unable to find a field that fits your needs among the standard fields, you have the option to create a new field by clicking the provided link.

6. Name your field.

7. Choose a field type from the dropdown.

8. Click **Create Field** to save your work.

9. Your new field will be automatically added to your form.

Editing and removing fields

Having the capability to effortlessly delete fields from a form, whether it is for internal or external use, is crucial, especially when cloning and subsequently modifying the cloned template. To remove or change a field on your form, follow these steps:

1. Click the ellipsis to the right of the field name.

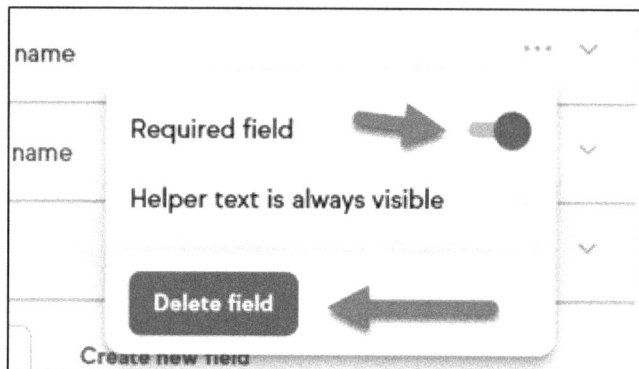

Figure 6.8 – Removing fields from an internal form

2. Use the toggle to make the field required or not required.
3. Set your helper text to always visible or not using the toggle.

To delete fields, follow these steps:

1. Click the ellipsis to open the menu.
2. Click the **Delete** button.

To move form fields, follow these steps:

1. Click and hold the field you want to move.
2. Drag the field up or down and then release the mouse button to place it above or below other fields.

Automating internal forms

Internal forms are great for task-driven processes. While a public form sends information based on what the potential customer may do, an internal form can drive what your team does. This can keep leads moving through your sales cycle when a more personal touch is needed.

For this next step, let's create a custom automation that will send a 48-hour follow-up email and create a task for our team.

1. Click the **Next** button or the **2** in the navigation to move to the **Automate** section.

2. Click the **+Create custom automation link**.

3. The automation builder will slide open from the right with a pre-named automation and the **When** field already set.

4. In the **Then** box, click the + to add your automation step.

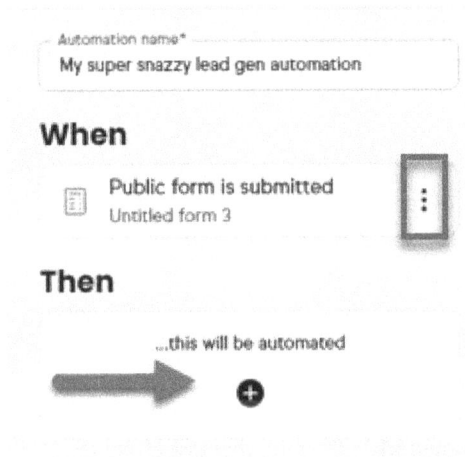

Figure 6.9 – Using the navigation bar to open the Automate feature

5. You now have several options for automating your workflow:

    A. **Communications**: Sends texts, pages, or notifications

    B. **Pipeline**: Moves a deal stage or creates a new deal

    C. **Tags**: Applies or removes a tag

    D. **Tasks**: Creates a task

6. Scroll to the bottom of the list and choose **Create a task**.

7. Give your task a title.

8. Add a description. Remember this is for your team so be specific with the details.

9. Set a due date for the data.

10. Assign the task to the contact owner, yourself, or someone on your team.

11. Click the **Next** button to continue.

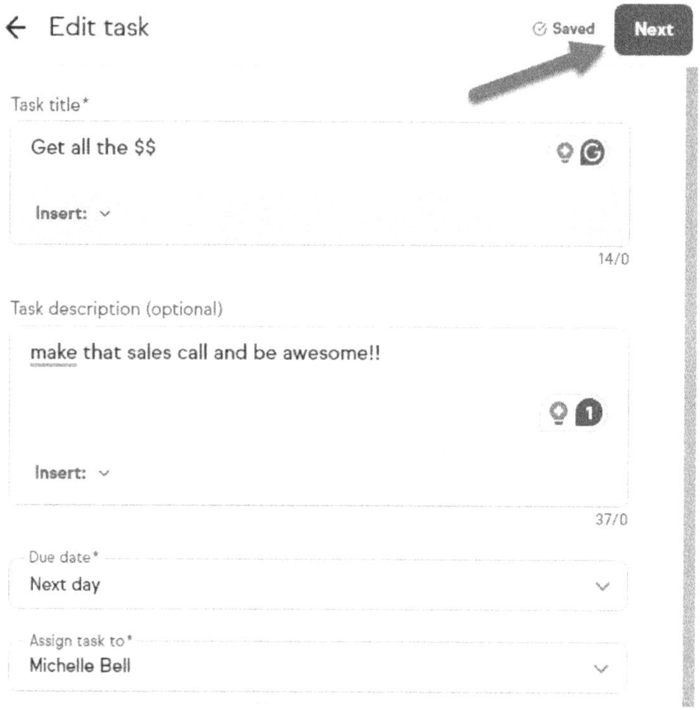

Figure 6.10 – Defining your task attributes

12. Choose **Immediately** to create your task right away or **Delayed** to schedule the task.

Publishing your internal form

Once your form has been built and you have added automations, you're ready to move on to publishing:

1. Click the **Close** button (in the lower left corner) to continue.
2. Click **Actions** in the top right corner and choose **Save and publish**.

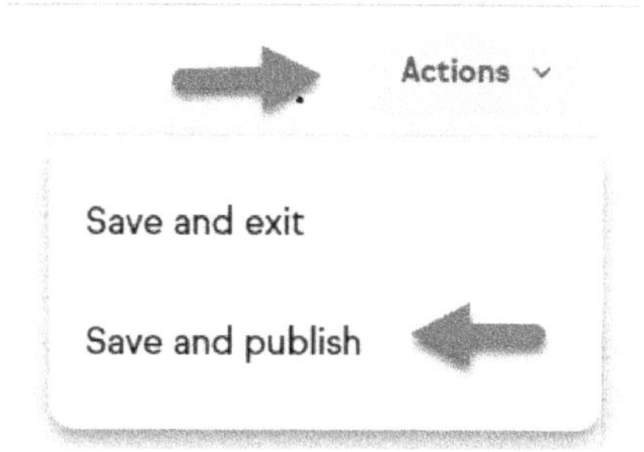

Figure 6.11 – Saving your work

Adding contacts using an internal form

You can use your form virtually anywhere in Keap. To use your internal form, follow these steps:

1. Click the + in the top left corner of the navigation pane.
2. Select **Contact**.
3. Click **Add a contact** to open the dropdown.
4. Choose your form. It is important to note here that only published forms will appear in the list.

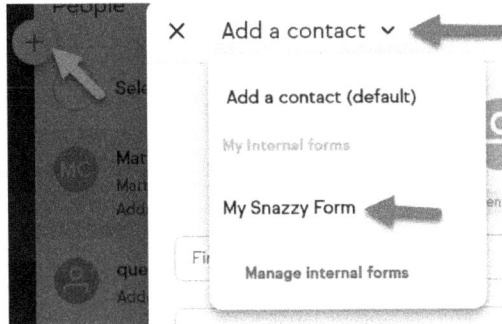

Figure 6.12 – Choosing your form to add contacts

Updating existing contacts with an internal form

1. Navigate to the desired contact record.
2. Click the **More** button.

Figure 6.13 – Updating an existing form

3. Select **Forms**.
4. Select your form.

5.   Complete the form.

6.   Click **Update contact**.

How it works...

Overall, using internal forms within a CRM provides organizations with a structured and efficient way to collect, manage, and utilize internal data, leading to improved collaboration, data quality, and workflow efficiency.

Landing pages

A landing page is a standalone web page designed with a specific objective in mind, such as prompting visitors to take a desired action, such as making a purchase, signing up for a newsletter, or downloading a resource. Unlike typical web pages, landing pages are crafted with minimal distractions to maximize conversions. They often feature compelling headlines, engaging content, and persuasive calls to action, and may include testimonials or social proof.

Landing pages play a crucial role in digital marketing campaigns by guiding visitors through a targeted conversion process, ultimately driving business goals and enhancing customer engagement.

Getting ready

Before you dive into crafting the *pièce de résistance* of marketing magic, be sure to gather up all the necessary ingredients. Skipping any of these essential elements might just leave your landing page feeling more like a parking lot:

- **Clear objective**: Define the purpose and goal of your landing page

- **Target audience**: Identify the audience you want to reach

- **Compelling headline**: Craft a headline that captures the attention of your audience and communicates the value proposition

- **Engaging content**: Write persuasive and concise content that resonates with your audience

- **Eye-catching visuals**: Use high-quality images, videos, or graphics to create visual appeal and emotional connection

- **Strong call to action (CTA)**: Include a clear and compelling CTA that prompts visitors to take the desired action

- **Minimal distractions**: Keep the design clean and uncluttered to focus attention on the main message and CTA

- **Mobile responsiveness**: Ensure that your landing page is optimized for mobile devices to reach users on all screens

How to do it...

Landing pages are structured into eight sections: content, blocks, body, images, uploads, audit, pages, and settings. One of the standout features of Keap is its ability to automatically save your work as you go, so you don't have to worry about getting it all done in one sitting.

1. Click on the **MARKETING** tab in the left-side navigation bar to open the menu and choose **Landing pages**.

2. Click the **Create a landing page** button.

3. You have several options to choose from:

    A. Start with a template from the gallery

    B. Choose one of your own templates

    C. Copy any previously published page

    D. Start from scratch

4. Name your landing page and click **Continue.**

Content

There are fourteen tools that we will be working with in the landing page builder. Each tool plays a significant role in how people see and interact with your landing page content. You will move between the canvas and the tool menu on the right as we build your page template.

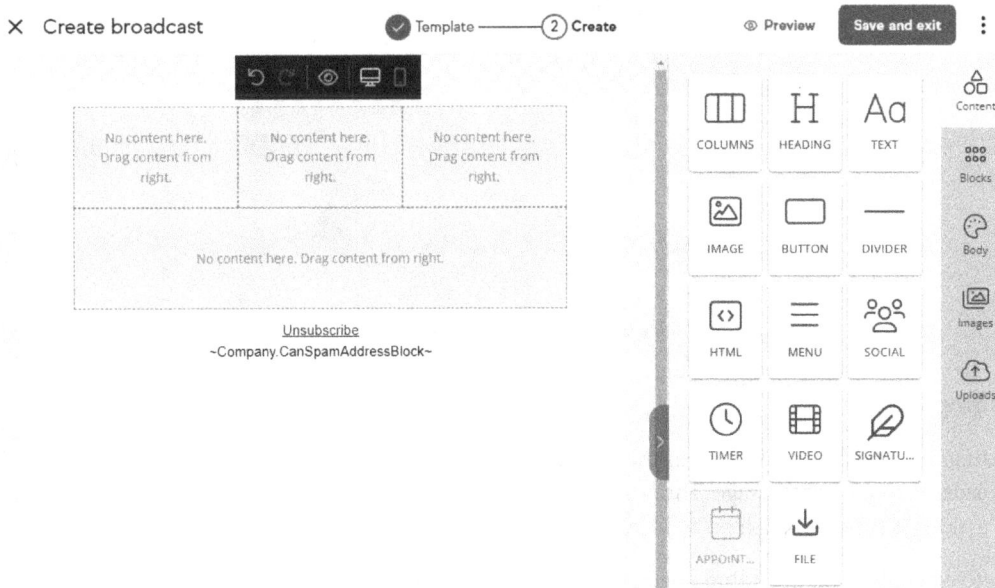

Figure 6.14 – Preview of the tools menu

Columns

Chapter 5 explored the intricacies of the email builder, thoroughly exploring its tools and features. It is worth noting that many of these concepts are also applicable to the form builder, as these tools share a significant portion of their features. This similarity simplifies the learning process when navigating through form and email templates.

As a refresher, columns play a vital role in styling content to achieve visual appeal. They enable the division of rows into multiple columns, facilitating the creation of side-by-side content. It is important to keep in mind that columns are most effective for content intended to be viewed on desktop devices.

To add columns to your page template, follow these steps:

1. Begin by clicking on and then dragging the **Columns** icon from the **Tools** menu onto the page canvas.

2. There are six column layout options for you to choose from. In the tools menu, click on the option you prefer.

3. Set a column background color if desired. Remember you can always add an image as the background for your row if you prefer.

4. Set the padding to push your content away from the edges of the column.

5. Set a section name. While this is not strictly necessary, doing so will save you a lot of time when it comes to updating formatting and optimizing copy placement.

6. As mentioned, columns work best in desktop view. You have the option to hide or force the columns to not stack by adjusting the toggle under **RESPONSIVE DESIGN**.

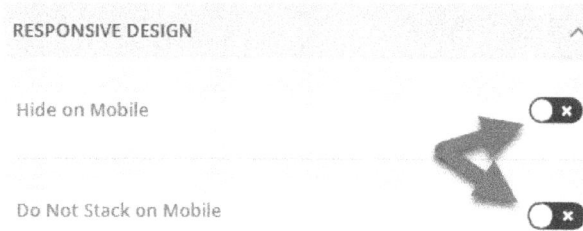

Figure 6.15 – Toggle buttons

Heading

Headings on a landing page serve as the titles or main sections that structure the content and guide the reader through the message. These headings are typically large, bold, or stylized text that stands out, making it easy for recipients to identify and navigate to different parts of the page.

Effective headings on a landing page template can enhance readability, drawing attention to key information and helping the user to take immediate action.

To add a heading to your page, follow these steps:

1. Begin by dragging the **Heading** icon from the **Tools** menu onto the page canvas.
2. Choose your heading size by clicking on **H1**, **H2**, **H3**, or **H4** .
3. Choose your preferred font from the drop-down menu.
4. Choose your preferred font weight from the drop-down menu.
5. Set the font size by typing in the box or using the +/- buttons.
6. Align your header by clicking on the left, center, right, or justified button.
7. You can set the line height by typing in a percentage or using the +/- buttons.
8. Use the **Responsive Design** toggles to hide or display your heading on mobile and desktop devices.

> **Note**
> When adding sections (rows), it is highly recommended that you name each section. As you build more robust landing pages and forms, the ability to easily move sections up or down relies on knowing which one you are moving. Without section names, this becomes a major hassle.

Text

The text on your landing page template should be clear, concise, and tailored to the intended audience. It should reinforce the message of the header and persuade the reader to keep going.

Proper formatting, including font styles, sizes, and colors, is essential for ensuring readability and visual appeal. Well-crafted text is crucial for effective communication, allowing the reader to immerse themselves in your story.

To add text to your page, follow these steps:

1. Begin by dragging the **Text** icon from the **Tools** menu onto the page canvas.
2. Choose your preferred font from the drop-down menu.
3. Choose your preferred font weight from the drop-down menu.
4. Set the font size by typing in the box or using the +/- buttons.
5. Select a font color by clicking on the color box to open the color editor:

    I. Click in the color area to free select.

    II. Type in your hex code in the provided hex box.

    III. Type in your RGB color codes in the provided RGB boxes.

6. Align your text by clicking on the left, center, right, or justified button.
7. You can set the line height by typing in a percentage or using the +/- buttons.

8. If you are adding links, you can choose to allow your links to inherit the body style of the section or use the toggle to turn off this feature and manually define the link attributes.

9. Use the general settings to change the container size for your text. The **more options** toggle will give you even more options for creating padding around your text. This is also where you will find the section name box. A section name is essential if you want to use links that point to a specific area of your landing page.

10. Use the **Responsive Design** toggles to hide or display your heading on mobile and desktop devices.

Images

Utilizing images on your landing page enhances the visual appeal and reinforces the messaging conveyed through text. These visual elements, such as photos, graphs, logos, and icons, play a crucial role in capturing the audience's attention, conveying information effectively, and evoking emotions.

Alt text, also known as alternative text, is essential for images on web pages, providing a concise description for users who may have images disabled or who rely on screen readers. By including relevant and compelling images, you can strengthen branding, increase engagement, and improve the conversion rates on your landing page.

To add images to your page, follow these steps:

1. Begin by dragging the **Image** icon from the **Tools** menu onto the page canvas.

2. There are several ways to select an image for your page:

    A. Search previously uploaded images

    B. Choose from stock photos

    C. Drop new images onto the box

    D. Insert the page using a URL

3. You can set the width of your image manually by toggling off the **Auto Width** button.

4. Align your image by clicking on the left, center, right, or justified button.

5. Add your **Alternative Text** in the provided box.

6. Set an action for your image by selecting one from the drop-down menu:

    A. Open Website

        i. Add the URL for the website in the URL box.

    B. Send Page

        i. Add the mail to address in the Mail to box.

        ii. Add your subject to the subject box.

    iii. Add any body text to the Body box.

  C. Call Phone Number

    i. Add the number you want to be dialed into the Phone box.

7. As previously noted, the ability to control the padding is found in the general section of the editor.

8. Use the **Responsive Design** toggles to hide or display your text on mobile and desktop devices.

Buttons

Buttons are the holy grail of landing pages! They are your calls to action and encourage people to click through to your website, opt in for an e-book or webinar, or, even better, make a purchase.

Buttons should be styled to stand out, using contrasting colors and high visibility.

To add a button to your landing page, follow these steps:

1. Begin by dragging the **Button** onto the page canvas.

2. Set an action for your button by selecting one from the drop-down menu.

    I. Open the website:

      i. Add the URL for the website to the URL box.

      ii. Choose open in the same tab or a new tab.

    II. Go to the page section. Using the dropdown, select the section you want to send the view to.

    III. Go to the landing page. Select another landing page that you have published.

    IV. Go to the landing sub-page. Select another landing sub-page that you have published.

    V. Go to the booking page. Add the URL for a booking or appointment page.

    VI. Go to the checkout form. Send the user to a published checkout form or create a new one.

3. Set your button's text color and background color by clicking on the respective boxes, and then follow these steps:

    I. Click in the color area to free select.

    II. Type in your hex code in the provided hex box.

    III. Type in your RGB color codes in the provided RGB boxes.

4. Use the design toggle to turn on/off Auto Width.

  A. **On**: The button will automatically change size to suit its container.

  B. **Off**: Uses the slider to set the button percentage to container.

5. Choose your preferred font from the drop-down menu.

6. Choose your preferred font weight from the drop-down menu.

7. Set the font size by typing in the box or using the +/- buttons.

8. Align your text by clicking on the left, center, right, or justified button.

9. You can set the line height by typing in a percentage or using the +/- buttons.

10. Button padding can be set to the same for all sides or customized on each side by using the toggle button.

11. Next, we need to set the button borders:

    I.   Choose your border style and set the pixel size. There are two options for border style:

        • Dotted ....

        • Dashed -------

    II.   Set the button's border text color and background color by clicking on the respective boxes. Then, follow these steps:

        i.   Click in the color area to free select.

        ii.   Type in your hex code in the provided hex box.

        iii.   Type in your RGB color codes in the provided RGB boxes.

    III.   The button border can be set to the same for all sides or customized on each side by using the toggle buttons.

12. Use the **Responsive Design** toggles to hide or display your button on mobile and desktop devices.

Dividers

Dividers are horizontal lines used to visually separate different sections of content within the page builder. These dividers help to create a clear and organized layout, preventing the page from appearing cluttered and making it easier for recipients to navigate through the information.

By breaking up the content into distinct sections, dividers contribute to a more aesthetically pleasing and user-friendly page design. They are especially useful in long-form story pages.

To add a divider to your page, follow these steps:

1. Begin by dragging the **Divider** icon onto the page canvas.

2. Use the slider to set the width of your divider.

3. Use the dropdown menu to choose your line type. There are three options:

    A. Solid _____

    B. Dotted ………………..

    C. Dashed ----------------

4. Align your divider by clicking on the left, center, right, or justified button.

5. Use the Responsive Design toggles to hide or display your heading on mobile and desktop devices.

HTML

HTML is commonly used to design and customize page templates when the onboard tools do not provide enough customization to suit your needs. For more info on this, check out *Chapter 4, Creating Content*.

To add HTML to your page, follow these steps:

1. Begin by dragging the **HTML** icon onto the page canvas.

2. Paste your HTML code into the provided box

3. Use the Responsive Design toggles to hide or display your HTML on mobile and desktop devices.

Menu

The purpose of a menu on a landing page is to provide navigation options for visitors to explore additional content or pages on the website. While the primary goal of a landing page is to encourage visitors to take a specific action, such as making a purchase or signing up for a newsletter, a menu allows users to access other relevant information or resources if they are not ready to convert immediately.

To add a menu to your page, follow these steps:

1. Begin by dragging the **Menu** icon onto the page canvas.

2. Click +**Add New Item** to begin adding your items.

3. In the action dropdown, choose your desired outcome:

    I. Open Website – sends the user to a specific page that you define.

        i. Add the URL for the website in the URL box.

        ii. Choose open in same tab or new tab.

    II. Go to page section – sends user to a specific section of your landing page.

        i. Using the dropdown select the section you want to send the viewer to. This is when having named your sections really comes in handy!

      III.   Go to landing page – sends the user to a landing page you choose.

        i.   Select another landing page that you have published.

      IV.   Go to landing sub-page – sends the user to a secondary page, as a thank you or upsell for example.

        i.   Select another landing sub-page that you have published.

      V.   Go to booking page – send the user to a booking link that you specify.

        i.   Sends viewer to your booking or appointment link.

      VI.   Go to checkout form – send the user someplace they can give you money!

        i.   Send user to a published check out form or create a new one.

4. Repeat *steps 2* and *3* until you have added all your menu options.
5. Choose your preferred font from the dropdown menu.
6. Choose your preferred font weight from the dropdown menu.
7. Set the font size by typing in the box or using the +/- buttons.
8. Select a text color by clicking on the color box to open the color editor:

    A.   Click in the color area to free select.

    B.   Type in your hex code in the provided hex box.

    C.   Type in your RGB color codes in the provided RGB boxes.

9. Select a link color using the same method used in *step 8*.
10. Align your text by clicking on the left, center, right, or justified button.
11. Choose a horizontal or vertical layout from the drop-down box.
12. You can add a separator between menu items by inserting a symbol (| or :) in the separator box.
13. Use the Responsive Design toggles to hide or display your menu on mobile and desktop devices.

Socials

Social icons on a landing page allow recipients to connect with you or share the page content on social networks. These icons typically include well-known platforms such as Facebook, Twitter, Instagram, and LinkedIn.

To add social icons to your page, follow these steps:

1. Begin by dragging the **Socials** icon onto the page canvas

2. There are three types of icons to choose from with three color options. On the toolbar, click the Icon Type dropdown and select from the following:

    A. Circle – full color

    B. Circle Black – black background with white image

    C. Circle white – white background with gray image

    D. Round (square with rounded edges) – full color

    E. Round (square with rounded edges) – black background with white image

    F. Round (square with rounded edges) – white background with gray image

    G. Square – full color

    H. Square – black background with white image

    I. Square – white background with gray image

3. Align your icons by clicking on the left, center, right, or justified button.

4. Set icon spacing by typing in the box or using the +/- buttons.

5. Use the Responsive Design toggles to hide or display your socials on mobile and desktop devices.

Timers

Dynamic countdown timers are embedded on landing pages to create a sense of urgency. They display the remaining time for a specific offer, promotion, or event and encourage recipients to take immediate action.

By using timers, you can capture the attention of your viewer, driving engagement and prompting them to make quicker decisions.

To add a timer to your page, follow these steps:

1. Begin by dragging the **Timer** icon onto the page canvas.

2. To set your **End Time** click in the box to open the calendar pop-up:

    I. Use the < > buttons to navigate to the month and year and select your date.

    II. Choose a time for your offer to expire on that date.

3. Choose a time zone for your offer from the drop-down box.

4. Set your preferred language using the drop-down box.

5. Select your background, digits, and label colors by clicking on the appropriate color box to open the color editor:

    A. Click in the color area to free select.

    B. Type in your hex code in the provided hex box.

    C. Type in your RGB color codes in the provided RGB boxes.

6. Choose your preferred digit font from the drop-down menu.

7. Choose your preferred label font from the drop-down menu.

8. You can set the width of your timer manually by toggling off the **Auto Width** button.

9. Align your timer by clicking on the left, center, right, or justified button.

10. Add your **Alternate Text** to the box.

11. You can set an action if someone clicks on the timer by choosing an option from the image link drop-down menu:

    I.   Open Website: Add the URL for the website in the URL box.

    II. Send Page:

         i.   Add the email address to the **Mail to** box.

         ii. Add your subject to the **Subject** box.

         iii. Add any body text to the **Body** box.

    III. Call Phone Number: Add the number you want to be dialed into the **Phone** box.

12. Use the **Responsive Design** toggles to hide or display your timer on mobile and desktop devices.

Videos

Integrating video into your landing page not only enhances engagement but also fosters authenticity and creates an emotional connection with your audience. Whether it is a heartfelt customer testimonial, a glimpse behind the scenes, or a storytelling narrative, video adds a human touch to your brand's message. By showcasing real people, experiences, or stories, videos have the power to resonate with viewers on a deeper level, evoking emotions and building trust.

This helps to strengthen your brand's credibility and relatability, ultimately driving conversions by establishing a genuine connection with your audience.

To add a video to your page, follow these steps:

1. Begin by dragging the **Video** icon onto the page canvas.

2. Add your YouTube or Vimeo URL to the **URL** box.

> **Note**
>
> The **Video** icon only works with YouTube and Vimeo

3. Set a padding to push your video away from the edges of the page

4. Use the **Responsive Design** toggle to hide or display your video on mobile and desktop devices.

Forms

In addition to video content, including forms on your landing page provides a valuable opportunity to capture leads and gather customer data. Forms serve as an interactive element that encourages visitors to take action, such as signing up for a newsletter, requesting more information, or downloading a resource. By strategically placing forms on your landing page, you can facilitate direct communication with your audience, nurture relationships, and guide them further along the conversion funnel.

Moreover, the information collected through forms provides valuable insights into user behavior and preferences, enabling you to tailor your marketing efforts and enhance overall campaign effectiveness.

To add a form to your page, follow these steps:

1. Begin by dragging the **Form** icon onto the page canvas.

2. Forms by default start with first name and email address. To customize your form, follow this step:

    I.   Give your form a name. This is a critical step that will save you time and frustration. When you leave the form with a default or general name it can be confusing to find the right form when you need to reference it later.

    II. The **Virtual field** box is an alternative to custom fields. These are great for triggering automation from a tag on a field when you only need this information temporarily. Data stored in virtual fields does not get stored in the contact record permanently.

    III. Use the slider to choose the width of your form.

    IV. Set your form alignment by clicking on the left, center, right, or justified buttons.

    V.   Set your label font, size, and color.

3. Set an action for your button by selecting one from the drop-down menu:

    I.   Open Website:

        i. Add the URL for the website in the URL box.

        ii. Choose open in same tab or new tab.

    II. Go to page section.

        i. Using the dropdown select the section you want to send the view to.

    III.   Go to landing page.

      i.   Select another landing page that you've published.

    IV.   Go to landing sub-page.

      i.   Select another landing sub-page that you've published.

    V.   Go to booking page.

      i.   Sends viewer to your booking or appointment page.

    VI.   Go to checkout form.

      i.   Send user to a published check out form or create a new one.

4. You can apply a tag when someone clicks the button by selecting it from the drop-down menu.

5. Use the **Responsive Design** toggle to hide or display your signature on mobile and desktop devices.

Checkout form

The Checkout form on a landing page serves as the final step of the conversion process, enabling visitors to complete a transaction or take a specific action, such as making a purchase or signing up for a service. It typically includes fields for users to enter payment information, shipping details (if applicable), and any additional necessary information to finalize the transaction. A well-designed checkout form should be streamlined, user-friendly, and secure, guiding users smoothly through the process and minimizing friction to maximize conversions.

To add an appointment link to your page, follow these steps:

1. Begin by dragging the **Checkout** icon onto the page canvas.

2. If you have already created a checkout form, you can choose it from the drop-down menu or create a new one.

3. Adjust the width of your form using the slider.

4. Align your form ensuring it flows smoothly with your previous section.

Appointments

Keap's built-in appointment booking feature allows you to embed a booking page directly into your landing page. Doing so not only streamlines the booking process but also empowers your visitors to take control of the process. And let's talk about the value of discovery calls! They are not just chats; they are excellent opportunities to truly connect with potential clients, understand their needs, and showcase how your programs or services can solve their problems. By offering this seamless functionality, you are not just enhancing the user experience—you are paving the way for meaningful engagements that can turn leads into loyal customers.

Blocks

Blocks are rows of prebuilt content you can drag onto the canvas saving you time. These pre-designed and customizable content elements or sections help you quickly create a visually appealing and structured design.

Once you have added blocks to your landing page you can arrange, customize, and edit them by going back to the content section. Blocks help you streamline page creation, offering a convenient way for you to craft visually engaging and cohesive landing pages.

To attach a block to your landing page, follow these steps:

1. To begin, click the **Blocks** icon in the menu bar on the right.

2. You now have the option to do the following:

    A. Search for a predesigned block by style (holiday, event, etc.)

    B. Add a blank block

3. Clicking the **all** link next to a style will filter the list for all blocks similar to that style.

4. Continue by dragging your selected block onto your canvas.

The content editor will automatically open so that you can customize your block. To add more blocks, simply close the content editor by clicking the **X** in the top right corner of the menu bar.

Body

Body is where you will configure the style of your overall landing page. You can set values for text, colors, and formatting that apply to your whole page. Body is broken down into three sections:

- The first section is General. This is where you will edit the basic look and feel of your landing page:

    A. **Text Color**: Use the color mixer to select or set the color of the text on your landing page

    B. **Background Color**: Use the color mixer to select or set the color of the page background

    C. **Content Width**: Use the slider to set the overall width of your content on the page

    D. **Font Family**: Use the drop-down menu to select your preferred font

    E. **Font Weight**: Use the drop-down menu to choose regular or bold text

- The second section is Links. This is where you will set the color and style of all the links on your landing page:

    A. **Color**: Use the color mixer to select or set the main color of all your links

    B. **Hover Color**: Use the color mixer to select or set the color people will see when visitors hold their mouse over any links on your page

C. **Underline**: Use the toggle to set whether your links will be underlined or not

D. **Hover Underline**: Use the toggle to set whether your links will be underlined or not when visitors hold their mouse over any links on your page

- The third section is Background. This is where you will upload any images you want to use as a background for your landing page. Images replace the background color when uploaded.

Images

Utilize the image feature to browse through a vast collection of images and elevate your landing page's visual appeal. Supported by Unsplash, Pexels, and Pixabay, you have access to millions of high-quality images. All images are licensed under Creative Commons Zero, ensuring unrestricted usage.

Uploads

Use the uploads block to import your images to Keap. Simply drag and drop your image or use the upload button to browse the images on your computer.

Audit

The audit tool provides a comprehensive examination of your landing page, ensuring its optimal functionality by scrutinizing the most crucial elements.

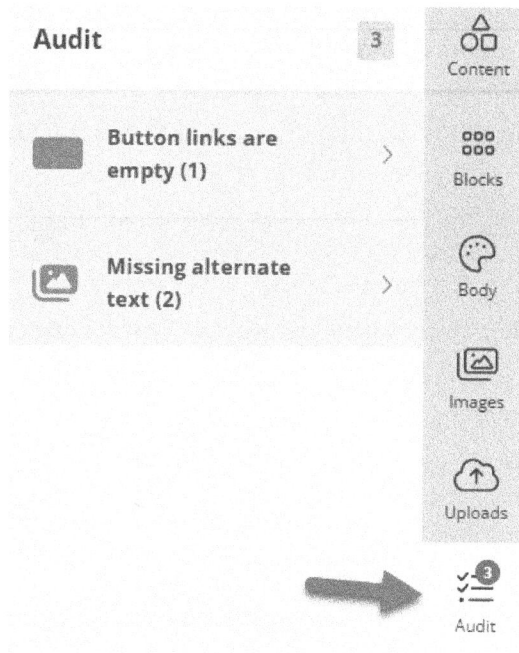

Figure 6.16 – Auditing your landing page

It meticulously checks for the following:

- Button links that are empty
- Missing image URL
- Missing alternate text

By conducting this assessment, the audit tool helps to rectify potential issues and ensures that your landing page is primed for maximum effectiveness and user engagement.

Pages

Use the **Pages** block to navigate between your landing page, any subpages, and your thank you page.

Settings

Use this block to enable Google and/or Facebook analytics.

Publishing your landing page

Before the world can view your snazzy new landing page, it will need to be published. Review your page and make any final edits.

To publish your landing page, follow these steps:

1. Click the **Next: Publish** button in the top right corner of the page.

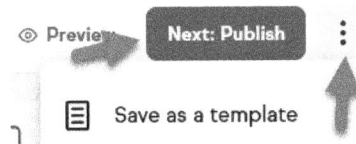

Figure 6.17 – Saving and publishing your landing page

2. Confirm you want to publish your page by clicking the **Publish landing page** button.
3. You now have three options for sharing your landing page:

   A. Use the hosted by Keap version by copying the URL.

   B. Copy the JavaScript snippet and place it on your website.

   C. Connect to a subdomain that you control. A subdomain is a prefix attached to a domain name to help separate a section of a website. This step requires you to add a CNAME record to your domain hosting account. To proceed, follow these steps:

      i. Click the **Edit URL** link above the copy URL button.

      ii. Next, click on the **Connect a subdomain** button.

iii. Type in the URL of your subdomain.

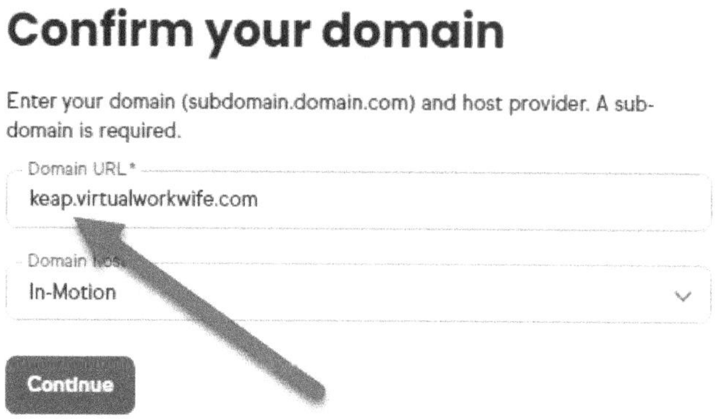

Confirm your domain

Enter your domain (subdomain.domain.com) and host provider. A subdomain is required.

Domain URL *
keap.virtualworkwife.com

Domain host
In-Motion   ⌄

Continue

Figure 6.18 – Adding your subdomain

iv. Choose your domain host from the drop-down menu or select **other** and then click **Continue**.

v. Keap will provide you with the values needed to create your CNAME record. It is highly recommended that you contact your domain hosting service for assistance with locating your CNAME records.

To save your landing page as a template, follow these steps:

1. Click the ellipsis to the right of the blue **Next: Publish** button.
2. Select the **Save as a template** option.
3. Name your template.
4. Click the **Save** button.

How it works...

Landing pages serve as focused entry points designed to convert visitors into leads or customers. They typically feature compelling content, persuasive calls to action, and engaging visuals to encourage specific actions. Elements such as forms and clickable links streamline user interaction, while integrated tools assess page performance and optimize user experience. By strategically combining these components, landing pages effectively capture attention, drive engagement, and facilitate conversions, ultimately supporting the overarching goals of marketing campaigns.

Part 4: Automation and Reporting

This part delves into automation techniques and reporting tools available in Keap, empowering you to automate repetitive tasks and gain valuable insights into your business performance.

- *Chapter 7, Easy Automations*
- *Chapter 8, Advanced Automations*
- *Chapter 9, Reports*

7

Easy Automations

With Keap's Easy Automations tool, you can easily create automated workflows and campaigns without needing any complex technical skills. With drag-and-drop technology and lots of prebuilt templates, you can quickly automate repetitive tasks, nurture leads, send targeted emails, and engage with customers at various stages in your pipeline.

When it comes to running a business, preset automations save time, increase efficiency, and deliver personalized experiences to your contacts. You can automate away any routine marketing and sales tasks, freeing you up to focus on high-value projects and create more growth in your business.

By mastering the Easy Automations tool, you'll unlock the power of efficiency and enhance your ability to engage with your audience effectively.

In this chapter, you will learn how to set up simple funnels that deliver content to your audience, apply tags based on their actions, and seamlessly move contacts in and out of pipeline stages.

We'll cover the following recipes in this chapter:

- Creating an Easy Automation from scratch
- Using a prebuilt automation template
- Editing automations
- Deactivating an automation
- Adding/removing contacts from an automation
- Deleting an automation
- Easy automation reports

Technical requirements

For this chapter, there are no technical prerequisites, so let's get started with automating your workflows and maximizing your productivity!

Working with Easy Automations

Before diving in, it's helpful to have a clear roadmap in mind. Consider creating a funnel map that outlines the contact's journey from the initial point of contact to becoming a repeat customer.

Once you have the basic outline, it will be easier to break up your automations into smaller groups or individual steps as you build them.

This strategic planning will provide a visual guide for designing automations that effectively guide people through each stage of the customer journey, ultimately driving them toward your desired goal faster and more efficiently.

How to do it...

Easy automations are designed to follow a *when, then, stop* cadence. For this recipe, we'll be creating an Easy Automation that says *when* a person completes a form, create a deal, *then* wait two days to create a task, and finally, *stop* if the deal stage moves to completed.

Creating an Easy Automation from scratch

1. Click on the **Automation** tab in the left-side navigation bar to open the menu, and then choose **Easy**.

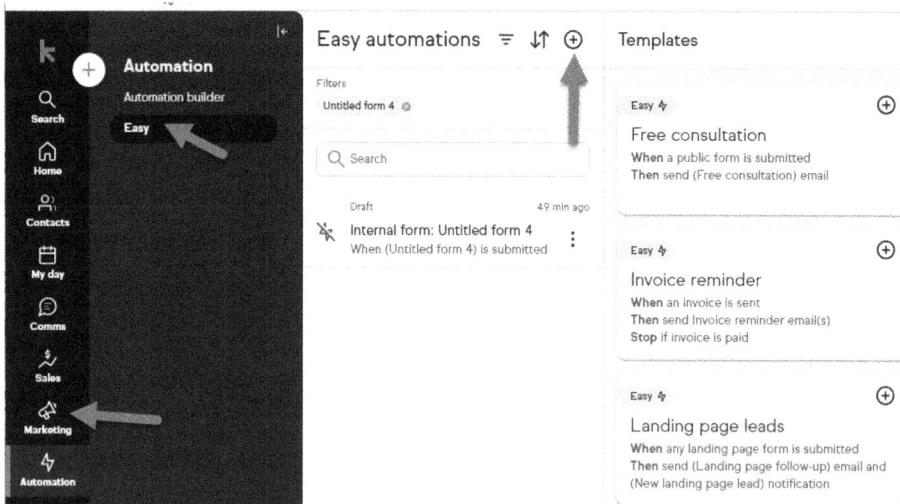

Figure 7.1 – Adding fields

2. There are several prebuilt Easy Automations to choose from. For this recipe, we will be starting from scratch by clicking the + symbol.

3. By default, your automation name will be **my automation** and the date. It is best if you rename your automation using something logical so that you can easily find it.

4. Click the **When** box. There are many options to choose from, and depending on which option you choose, you may need to take additional steps, such as selecting a tag, setting the appointment type, or which specific product will trigger your automation. To keep it simple, for this recipe, we are choosing **Public form is submitted**.

5. Locate your form by scrolling the list or typing the name in the box. You can make any necessary edits to the form by clicking the ellipses (...) to the right of the form name. Or you can create a new form on the fly by selecting **Create new form**.

> **Note**
>
> Each type of "when" will have a set of tools you can access by clicking on the ellipses (...), as seen in *Figure 7.2*. The most common tools are edit and the trash icon.

6. Click the **Next** button to continue.

7. We are now ready to add our "then" criteria. Click the **Then** box to choose an option.

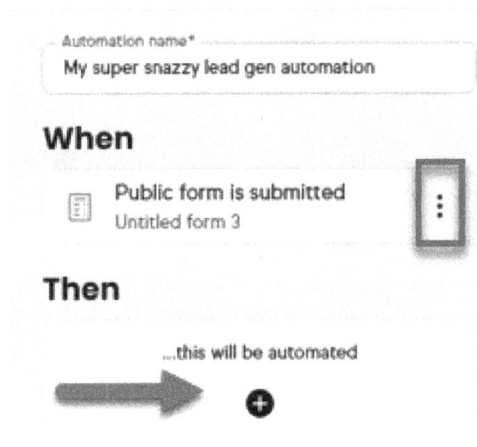

Figure 7.2 – Choosing your when and then attributes

8. Choose **Create a deal**.

9. Click the **Choose a stage** dropdown to select the pipeline you want to use for your deal and then select the stage you want to start in. For this example, let's choose **In progress**.

10. Give your deal a name. You can manually enter one or use the dropdown to select values from the contact record.

11. Give your deal a value.

12. Choose an owner from the drop-down menu.

13. Click **Next** to save your step.

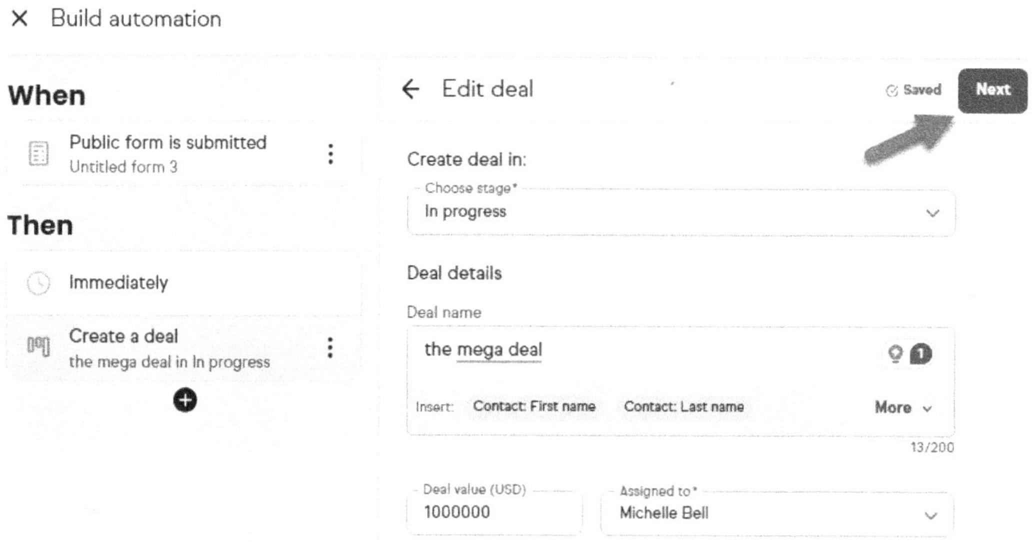

Figure 7.3 – Adding deal attributes

14. You can set your automation step to run immediately or choose **Delayed** to specify the number of hours, days, or minutes to wait before execution.

15. Click **Next** after making your selection.

16. Click the + symbol below your deal step to add another automation. This time, we'll choose **Create a task**.

17. Give your task a title.

18. Adding a description is optional, but highly recommended.

19. Set a due date for your task.

20. Use the drop-down menu to choose a person to complete the tasks.

21. Click **Next** to continue.

22. Choose a due date for your task, either **Immediately** or **Delayed**. For this recipe, we are choosing **Delayed**:

    A.   In the **Delay type** dropdown, select **Duration of time**.

    B.   In the **Delay duration** box, set the number of days to **2**.

23. Click **Next** to return to the automation builder.

24. Repeat *step 15* to continue adding automation steps.

Adding a stop rule to your automation

To add a stop rule, your automation needs to include at least one delayed then step. Adding a stop gives you the ability to kill any future automation steps when a contact meets a goal either by doing it themselves or by a user taking an action.

For example, you may want to stop any future emails asking them to make a purchase once they actually make the purchase.

1. To add criteria for stopping your automation, click the **Stop** box directly below your **Then** step.

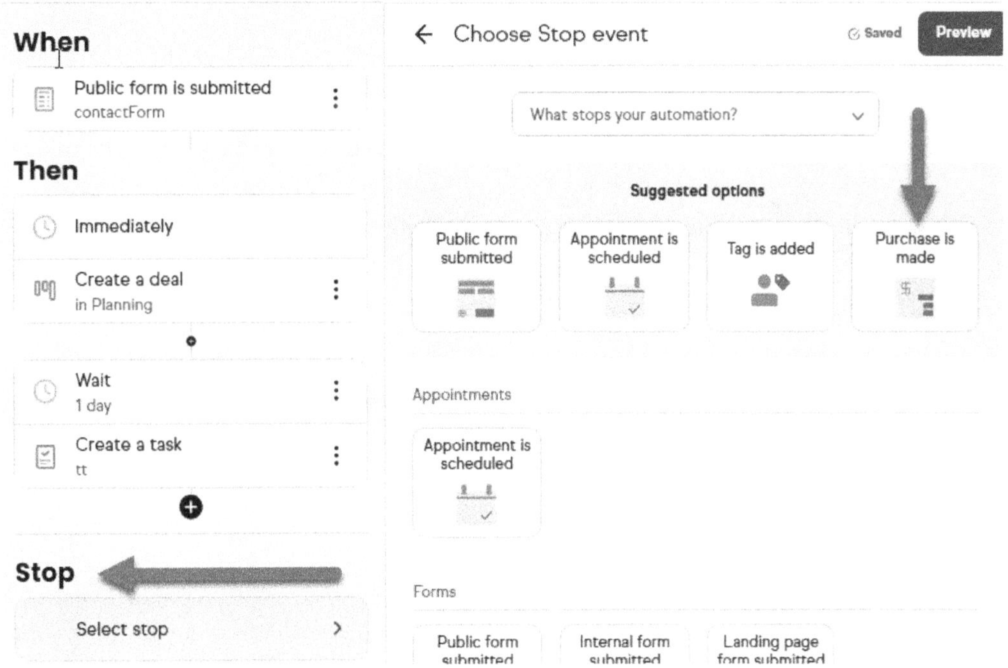

Figure 7.4 – Adding stop attributes

2. Choose **Purchase is made**.

3. You have three options for how a purchase can stop your automation:

   I.   Choose the **Any product or service** button.

   II.   Type the product name or scroll the list and select a product.

   III. Click the + **Create a new product or service** link.

4. For this exercise, we will be choosing **Any product or service**.

5. You will see either a **Preview** or **Publish** button in the top-right corner. The **Preview** button only appears when there is something that needs to be fixed.

6. Check your work and then click **Publish** to save your automation.

> **Note**
> As with tagging, having a standardized naming convention can greatly improve the ease of use of your CRM.

Using a prebuilt automation template

Your Keap CRM offers many prebuilt Easy Automation templates. Each template can be edited to add or remove steps, making them a fast and agile way to get your automations set up quickly.

1. Automations are saved by default with a generic name. Start by renaming your automation with something that indicates what actions the automation will run.

2. Click on the **Automation** tab in the left-side navigation bar to open the menu, and then choose **Easy**.

3. Click the plus button (+) on the template you want to use.

4. Edit the template as needed (refer to *steps 6-14* in *Creating an Easy Automation from scratch*).

5. You will see either a **Preview** or **Publish** button in the top-right corner. The **Preview** button only appears when there is something that needs to be fixed.

6. Click **Publish** to launch your automation.

Editing an Easy Automation

As your business grows, it may become necessary to edit your automations. You can add, remove, or edit the steps of your automation as needed. It is important to note that you must publish your automation after making edits in order for your edits to be deployed.

1. Locate the automation you want to edit by scrolling through the list.

2. Click the ellipsis on the right and choose **Edit**.

3. Clicking on a **When, Then,** or **Stop** box will open the item.

4. You can now edit the details of the automation step.

5. Click **Next** to continue.

6. You will see either a **Preview** or **Publish** button in the top-right corner. Remember, the **Preview** button will only appear when there is something that needs to be fixed.

7. Click **Publish** to relaunch your automation.

Deactivating an Easy Automation

When you turn off Easy Automation, new contacts won't join the automation, and any events triggered by contacts who previously met the when condition will stop. All contacts will be removed from the automation upon deactivation. You can reactivate the automation by making edits and republishing it.

1. Locate the automation you want to edit by scrolling through the list.

2. Click the ellipsis on the right and choose **Deactivate**.

3. A popup will appear to confirm that you want to deactivate your automation.

4. Click **Yes, deactivate** to continue.

Viewing and removing contacts in an Easy Automation

1. Locate a published automation by scrolling through the list.

2. Click the **Automation** box to open the automation details.

3. The number of active contacts currently in the automation is displayed. Click the number to view the contacts.

4. Check the **Select all** circle and click on the individual contacts you want to remove from the Easy Automation.

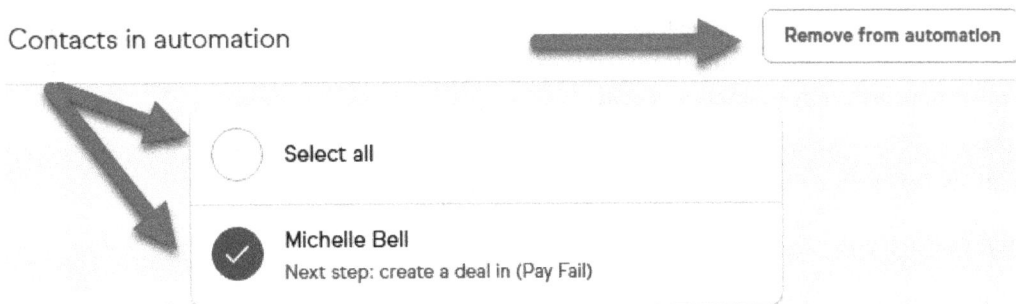

Figure 7.5 – Removing contacts from an easy automation

5. Click **Remove from automation**.

6. Click **Yes, remove** to confirm.

Deleting an Easy Automation

When you delete an automation, it's like tidying up your CRM space. It removes the automation entirely, stopping any future events for contacts who previously met the criteria. And remember, when you delete it, all contacts involved in that automation are removed as well, keeping things neat and tidy!

1. Locate a published automation by scrolling the list.

2. Click the ellipsis on the right and choose **Delete**.

3. A popup will appear to confirm that you want to delete your automation.

4. Click **Delete Automation** to continue.

Checking Easy Automation reports

It's easy to check the stats for your Easy Automations. Knowing what your automations are doing for you (emails/texts sent, tasks/deals created, appointments booked, etc.) is essential to understanding what efforts are maximizing conversions and where you may need to make changes in order to optimize your offerings.

1. Navigate to **Easy Automations**.

2. Locate a published automation by scrolling the list.

3. Click the box to open the details of the automation.

4. Click the **View Metrics** button at the top of the automation.

How it works...

Keap's Easy Automations offer a user-friendly solution, enabling businesses to automate tasks without requiring advanced technical skills. This section explores how Easy Automations work and how they can revolutionize the way businesses operate.

Easy Automations operate on a simple yet effective framework consisting of when, then, and stop conditions. Let's look at each of them.

When conditions

Easy Automations start with a when condition typically triggered by a lead or client action. Here is a list of common when conditions:

- An appointment is scheduled

- An appointment is cancelled

- A public form is submitted

- A landing page form is submitted

- An internal form Is submitted

- A specific product is purchased

- Any product is purchased

- An invoice is sent

- An invoice is paid

- A quote is sent

- A quote is accepted

- Checkout form is paid

- Deal enters stage

- Deal exits stage

- Deal is closed

- A tag is added to a contact

Then conditions

After a contact enters your automation, you have the option to set up one or multiple events to occur. Here are some typical "then" conditions you can configure:

- Send an email to the contact

- Send a notification (desktop, mobile, email) to a specific Keap user in your CRM

- Send a notification (desktop, mobile, email) to all Keap users in your CRM

- Send a notification (desktop, mobile, email) to the contact's owner

- Add a tag to the contact

- Remove a tag from the contact

- Create a task and assign it to the contact's owner

- Create a task and assign it to a specific Keap user in your CRM

- Create a Pipeline deal

- Move a deal

> **Note**
> Mobile notifications require the Keap user to have installed the Keap Mobile app.

Stop conditions

When contacts meet stop conditions, you have the ability to remove them from an Easy Automation. To incorporate stop conditions into your automation, it's necessary to include at least one delayed then step. Here, you'll find a compilation of common stop conditions:

- A new appointment is scheduled.

- A public or landing page form is submitted.

- An internal form Is submitted.

- A specific product is purchased.

- Any product is purchased.

- A quote is accepted – This option will only appear if the WHEN automation is set to **Quote Sent**. It does not appear in the regular list.

- A quote is sent.

- An appointment is canceled – This option will only appear if the WHEN automation is set to **appointment is scheduled**. It does not appear in the regular list.

- An invoice or check out form is paid.

- Deal enters or exits a stage.

- Deal is closed.

- A tag is added to a contact.

With Keap's Easy Automations, you can effortlessly automate repetitive tasks, nurture leads, and connect with customers seamlessly. Plus, the user-friendly interface makes it a breeze for anyone to use; no complex coding is required!

Advanced Automations

The **advanced automation builder** is incredibly versatile, offering a whole toolbox of functionalities to tackle even the trickiest automation challenges.

We just learned about easy automations and their "when," "then," and "stop" approach. Those concepts are also true for advanced automations, but we're going to be using a more agile builder – one where we can incorporate multiple logical steps so that we can create pathways as limitless as our imagination. In the advanced automation builder, your "when" and "stop" are goals, while sequences are your delivery mechanism for your "then" actions.

In this chapter, we'll cover the following recipes:

- Connecting elements
- Versioning
- Understanding goals
- Understanding sequences
- Working with decision diamonds
- Building an advanced automation

Technical requirements

For this chapter, it would be helpful, but not required, if you have a general idea for a lead capture funnel – for example, you may have an ebook or worksheet you want to share via social media. This would be a great option for the Building an advanced automation section of this recipe.

Using the advanced automation builder

Advanced automations are simply a series of connected goals and sequences. Goals are placed before or after a sequence. Once connected, a goal will stop sequences on the left (unless programmed not to) and start sequences on the right. In this way, a goal can perform multiple functions:

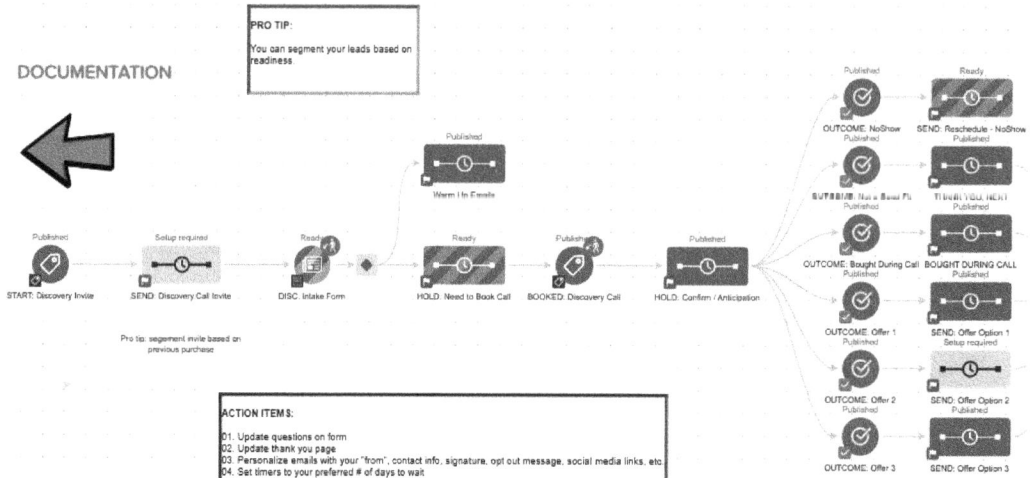

Figure 8.1 – Connected elements in an automation

Figure 8.1 shows an example of advanced automation with several goals and sequences and shows how they are connected. One of the most agile features of advanced automations is the ability to create a "one-to-many" scenario. In this example, there's one sequence with many outcomes or goals. We had one appointment with a potential client and there are many possible outcomes:

- They were a no-show
- They showed up and purchased
- They showed up but it wasn't a good fit
- They showed up but didn't buy and we need to send a follow-up offer

Depending on how many products or packages you sell, you may want to have several different automated offers so that your number of possible outcomes is infinite.

You cannot connect a goal directly to another goal. There must be a sequence between them:

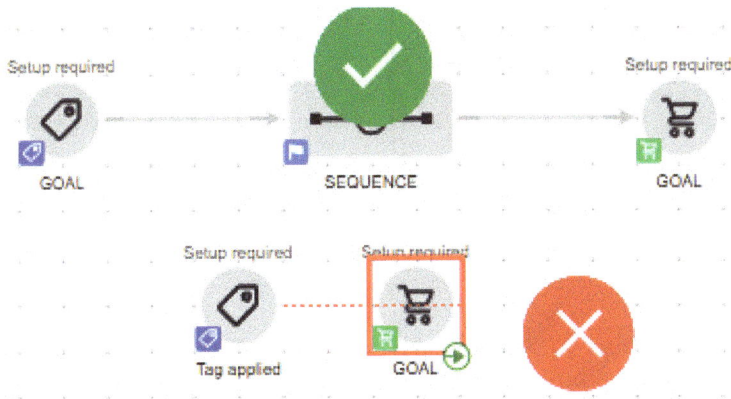

Figure 8.2 – Correct versus incorrect connections

> **Note**
>
> Contacts in advanced automation will always be pulled to the farthest goal met. For example, *Figure 8.1* shows a sequence inviting potential clients to an appointment to stop the invitations once they book. We already know the next step is a sequence where we are waiting for the outcome of the appointment, which triggers the follow-up to happen. If a person hasn't booked their appointment yet but purchases the product, they will be pulled past the outcome step because it is no longer needed. This will be important to note as you build more complex workflows.

How to do it...

In this recipe, we will learn how to navigate the advanced automation builder, its components, and how they work together to create workflows and funnels.

Navigating the canvas

Advanced automations can often become quite extensive, depending on the number of steps needed for your workflow. A hiring campaign, for example, may have 15 to 20 sequences as you move a candidate from pre-interview to handing them the keys to the office.

When you're working on a large automation and discover you need more canvas area to work with, simply click and hold your mouse inside the canvas and move your mouse to the left. This will move your entire automation and allow you to see more blank canvas areas.

Keyboard shortcuts

Keyboard shortcuts are a quick way to move within the canvas. You can find a list of shortcuts by locating the keyboard button in the lower left corner of the canvas:

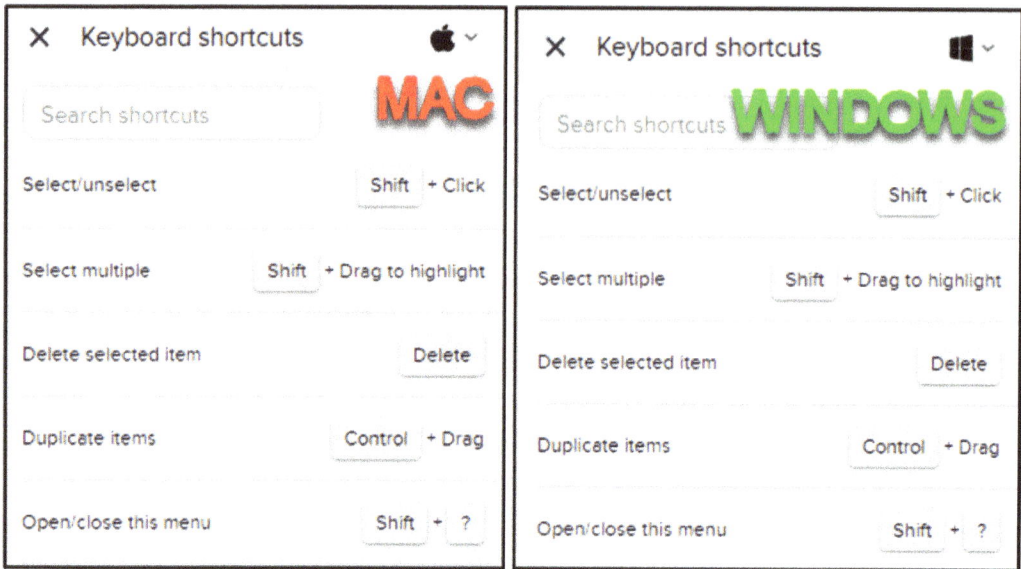

Figure 8.3 – Keyboard shortcuts

Moving multiple goals or sequences as a group

Follow these steps:

1. Hold down the *Shift* key and click each element you want to move.
2. Mouse over any of the selected elements.
3. Click and hold your mouse button, then drag the elements to another area of the canvas.
4. Release your mouse to drop them.
5. Click on any open area of the canvas to deselect the elements.

Deleting elements

To delete an element on the canvas, do the following:

1. Click on the item you want to delete and do one of the following:

    A. Choose **Delete** from the pop-up menu.
    B. Click *Delete* on your keyboard.

> **Warning!**
> When you delete a sequence from the canvas and then publish, any content contained within the sequence, such as notes or emails, will be deleted as well.

Renaming elements

Renaming your elements can save you time and effort when the time comes to change or update them. For example, emails are easier to locate if you rename the mail icon so that it matches your subject line.

There are two ways to rename elements in an automation:

1. Double-click the name of the element.

2. Right-click on the element and choose **Rename**.

> **Note**
>
> The only elements you can't rename are the apply/remove tag elements.

Connecting elements

Think of connecting elements as laying the groundwork for your workflows. It's like putting the first few pieces of a puzzle together, setting the stage for your automated processes to take shape.

To connect elements, you must do the following::

1. Hover over the goal or sequence you want to connect. A green arrow will appear.

2. Click and hold the green arrow, then drag it over the goal or sequence you want it to connect to. It will turn green to indicate a successful connection. *If the line turns red, you cannot connect to that element.*

3. Release your mouse to complete the connection:

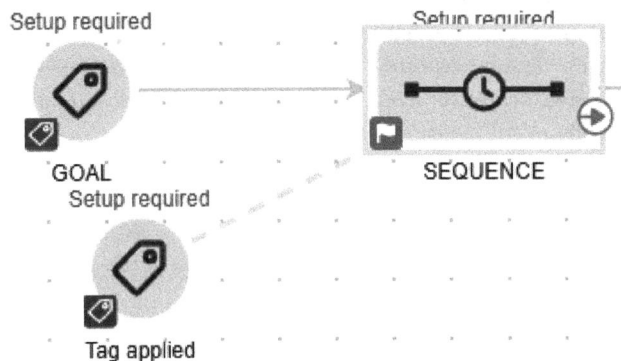

Figure 8.4 – Connecting elements

> **Note**
>
> You can insert an element in between two other elements by following these steps.

4. Grab an element from the toolbar and drag it over the connecting line between the existing elements.

5. When the line turns dark blue, release your mouse to insert the element:

Figure 8.5 – Solid connection

6. If the line is light blue, the element will *not* be connected, even though it may appear to be in the right place.

7. To verify that your element is connected, click on it and drag it above or below the line. You will immediately see the line either move with the element or not.

Deleting a connection

You can disconnect elements by simply clicking on the line between them and hitting the *Delete* key on your keyboard.

Accessing automation version history

The advanced automation builder has your back with an auto-save feature that kicks in every 30 seconds, guaranteeing you won't lose your hard work. You can also manually save a version if you plan to make significant changes. Versions capture the essence of your **automation** – think traffic sources, goals, sequences, and decisions. However, keep in mind that they do not store previous content, such as landing pages, email copy, or the decision diamond settings.

How to do it...

In this recipe, we'll cover how to create, access, and restore versions, a critical function for maintaining and optimizing your advanced automations.

Restore

Versions are date-stamped and stored so that you can roll back an automation to a previous copy. This is especially handy if you want to undo changes that are failing to perform. However, it's important to consider what might happen to contacts who are in your automation if you do a rollback. They may move unexpectedly or be removed completely from the automation. It's always a good idea to save the current version of the automation before you restore a previous version, just in case you change your mind!

Follow these steps:

1.   Click on the **AUTOMATION** tab on the left-hand side navigation bar to open the menu and choose **ADVANCED**.

2.   Locate the automation you want to edit and click the ellipsis on the right.

3.   Click **Finish setup** to enter the automation.

4.   Click the **Actions** button in the top-right corner.

5.   Select **Restore**.

6.   Click on the date stamp for the version you want to restore:

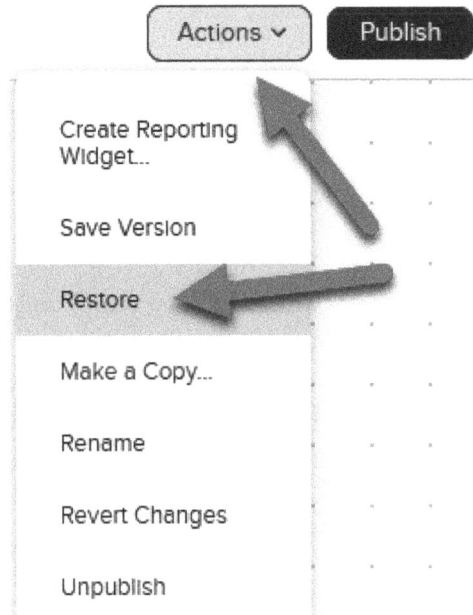

Figure 8.6 – The Actions menu

Revert Changes

With the **Revert Changes** option, you can take your automation back to how it was when you first opened it during this work session. To do so, follow these steps:

1.   Click the **Actions** button in the top-right corner.

2.   Select **Revert Changes**.

Make a Copy...

Creating fresh advanced automation is simplified by the ability to duplicate an existing one. Any changes made to the duplicated animation won't impact the original version. By default, a duplicated automation is unpublished, giving you time to update any form code, external links, or landing page links or redefine your goals. Follow these steps:

1. Click on the **AUTOMATION** tab on the left-hand side navigation bar to open the menu and choose **ADVANCED**.

2. Locate the automation you want to edit and click the ellipsis on the right.

3. Click **Finish setup** to enter the automation.

4. Click the **Actions** button in the top-right corner.

5. Select **Make a Copy...**.

6. In the popup, fill in the **Name value** for your copy.

7. Click **Save**.

Understanding goals

Goals within an automation are usually the actions we want people (internal or external) to take. Whether it's filling out a form, making a purchase, or scheduling an appointment, these are all goals that guide contacts from your sales process to your fulfillment process. The following descriptions provide a quick overview of each goal you can use in an advanced automation.

Goals can be sorted into two categories: those completed by a contact and those completed by a Keap user.

Goals triggered by a contact

The following goals can be triggered by a contact:

- **Web Form Submitted**: This goal is used to build a form inside the advanced automation builder.

- **Landing Page**: A Keap-hosted web page that can be used to collect leads, sell products, or deliver content.

- **Email Link clicked**: If an email's purpose is to get the reader's attention, then the links you include are the icing on the cake. They are the markers that tell us how captivated your reader is and assess their level of interest. When selecting a link click as a goal, it is essential to tie it to a distinct call to action (for example, "contact me" or "sign me up").

- **Product purchased**: Sales transactions are monitored when a contact completes a purchase via a Keap order form or shopping cart. This goal can be set wide, to track "any" purchase, or narrow, to track only a specific product.

- **Failed Purchase**: Using this goal, you can create an abandoned cart flow.

- **Quote status**: The quote goal can be set to trigger when a quote has been sent, viewed, or accepted.

- **Web Page automation**: This goal is triggered when an existing contact interacts with your website.

- **Appointments**: This goal can be set to trigger when contact schedules, reschedules, or cancels an appointment that is scheduled via the Keap appointments tool.

- **Lead Score achieved**: Lead scoring allows you to set a value for each interaction a contact has with you. The lead score goal can get set to fire when a contact increases or decreases to a set lead score.

- **API**: Use this tool when you have a third-party application, and you want to it communicate with Keap and process incoming activity.

- **WordPress Opt-In**: Like **Web Form Submitted**, this goal is achieved when you have installed the Keap WordPress plugin on your website.

Goals triggered by a Keap user or automation

The following goals can be triggered by a Keap user or automation:

- **Tag applied**: Tags are searchable labels that are used to segment your contacts. They can be applied and/or removed automatically as part of a sequence or manually by a Keap user.

- **Form submitted**: Internal forms are flexible in that they give you the ability to submit data on behalf of someone else or to document your interactions with a contact. Internal forms are the boss when it comes to automating internal workflows.

- **Task completed**: Tasks that have been added to a sequence can be used to satisfy the task completed goal. Moreover, you can set specific task outcomes to drive contacts down different automated paths, depending on the outcome chosen.

- **Pipeline Stage moved**: Pipeline stages are milestones in your sales process and track prospect progress. You can use this goal to trigger follow-up activity when someone enters or exits a stage:

Goals

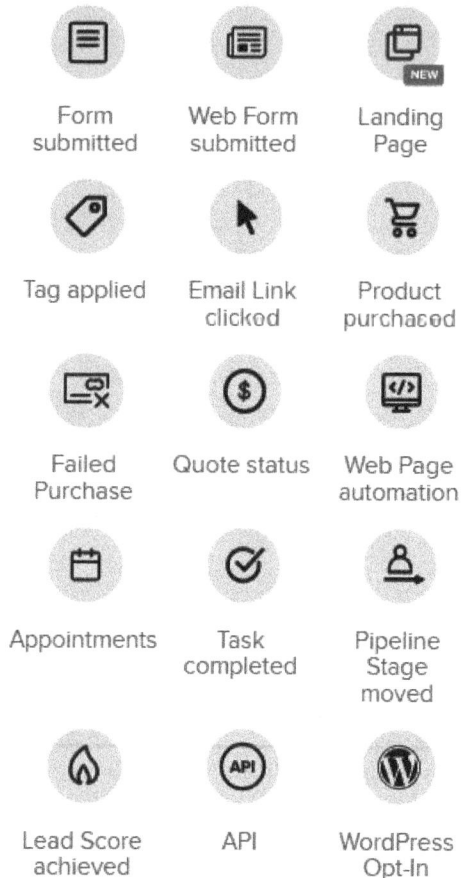

Figure 8.7 – Goal tools in advanced automations

Understanding sequences

Automation sequences are the meat and potatoes of advanced automations. They are where you will schedule a series of communications and/or processes to engage your contacts and drive them to move through your funnels. Most often, sequences are set in motion when an automation goal is achieved. Like a flowchart, individual sequences are small pieces of an overall larger automation strategy.

Just like automations, a sequence is created by adding drag-and-drop elements to a blank canvas. Once an element has been added to the canvas, you can edit and customize it. When an element is added to a sequence canvas, it automatically generates the related feature.

There are four main types of elements in a sequence: timers, communications, processes, and notes.

Timers

Timers are exactly what they sound: they control the time delay between triggered sequence elements. Let's look at four different types of timers:

- **Start Timer**: The start timer is a must-have and is already positioned on the canvas to signify the start of the sequence. If you need to run multiple communications or processes simultaneously, you can include more than one start timer. For instance, you might want to distinguish between emails sent to contacts and internal tasks:

Start

Figure 8.8 – Start Timer

- **Delay Timer**: Quite possibly the most often used timer, delay timers determine the time that passes between each step in your automation. Delay timers are excellent for "evergreen" content. They can do the following:

  A. Run after a set number of minutes, hours, days, weeks, months, or years

  B. Run on any day, weekday, weekend, day of the month, or day of the year

  C. Run at any time of day, specific time, or within a time range:

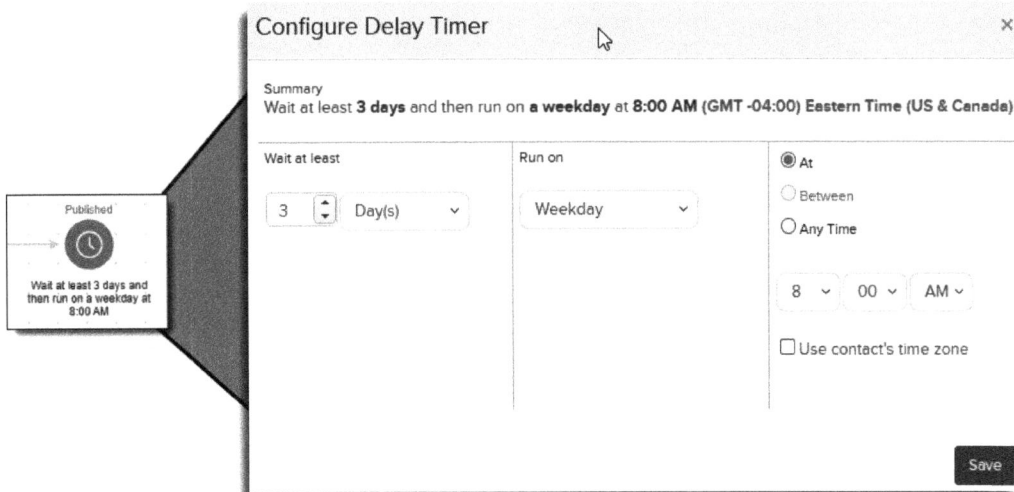

Figure 8.9 – Delay Timer

- **Date Timer**: Use the date timer when you want to set a specific calendar date and time to run your actions. These require care and attention as they need to be updated once the date passes. For example, if you have a timer set to send an email on a specific date, any contact who enters the sequence after that date will not receive the email:

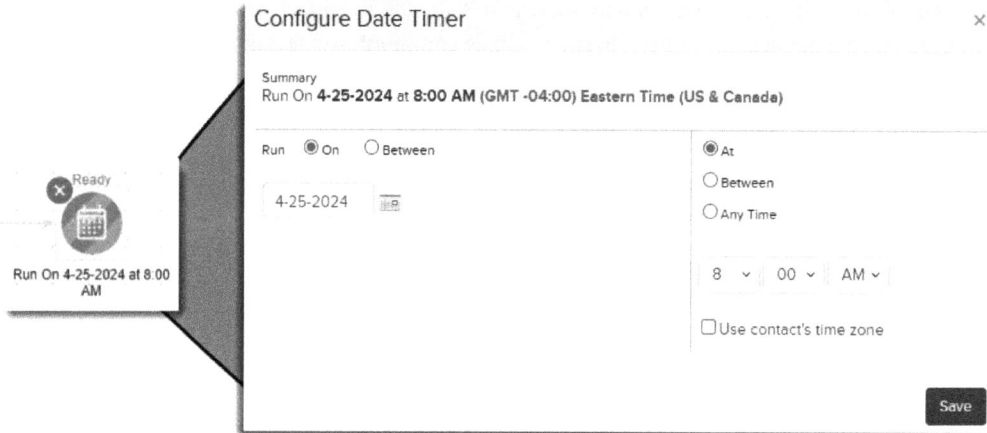

Figure 8.10 – Date Timer

- **Field Timer**: Field timers are based on a date field in the contact record, such as a contact's birthday or anniversary. If you've collected a custom date via a form field, you can use that date as a field timer:

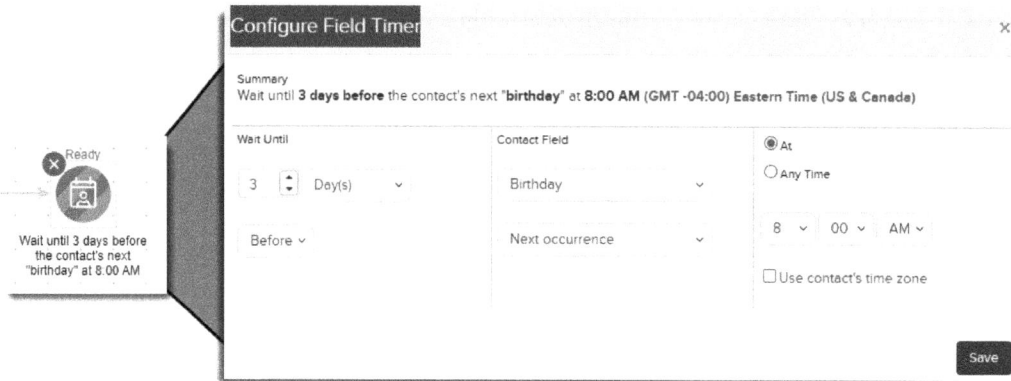

Figure 8.11 – Field Timer

> **Note**
>
> Pay close attention to the "occurrence" drop-down setting. If you select **Next occurrence**, the timer will use the month and date to schedule the event. Upon selecting the **Use Year from Field** option, the timer will use the month, date, and year to decide when to schedule the event.

- **Appointment Timer**: This timer is used when an appointment goal within the automation is met. Often, this is used to send email or text reminders before an appointment:

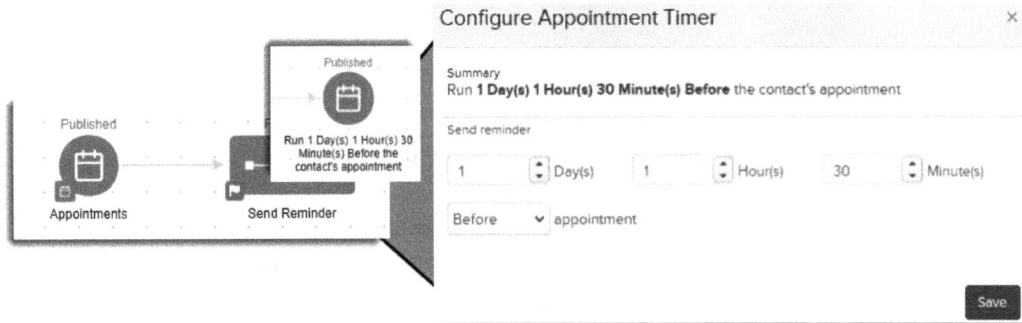

Figure 8.12 – Appointment Timer

Communications

The most common type of communication that's used in a sequence is emails. They are generally sent to a contact who is moving through your automation flow. You can also send text messages via an automation sequence:

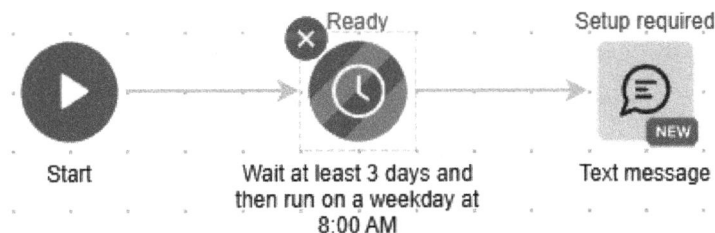

Figure 8.13 – Automating text messages

Please refer to *Chapter 4, Communicating with Your Lists*, for more information on setting up and sending text messages.

Processes

Process elements are all about organization, documentation, and keeping your workflow on track. Unlike communication elements, they don't directly send messages to contacts. Instead, they focus on updating contact records and assigning tasks to Keap users. Let's dive into the 10 types of process elements:

- **Apply/Remove Tag**: Tags serve as searchable labels that are utilized for segmenting contacts and monitoring specific contact interactions. Additionally, they are useful for notating the fulfillment of goals and creating reports that keep you on top of your automation strategy. We covered the basics of tagging in *Chapter 3, Managing Contacts*:

Figure 8.14 – Applying and removing tags with automation

- **Apply Note**: The note element adds a date-stamped note to a contact. You can add specific details to the note to document how and when a contact passed through your automation.

- **Create Task**: The task element is used to assign a Keap user a specific duty. Tasks are typically used when human intervention is needed, such as when updating a call outcome (attended, no-show, and so on) or sending a personal greeting card. Task completion can be used as a goal in the greater automation scheme, as seen in *Figure 8.1*:

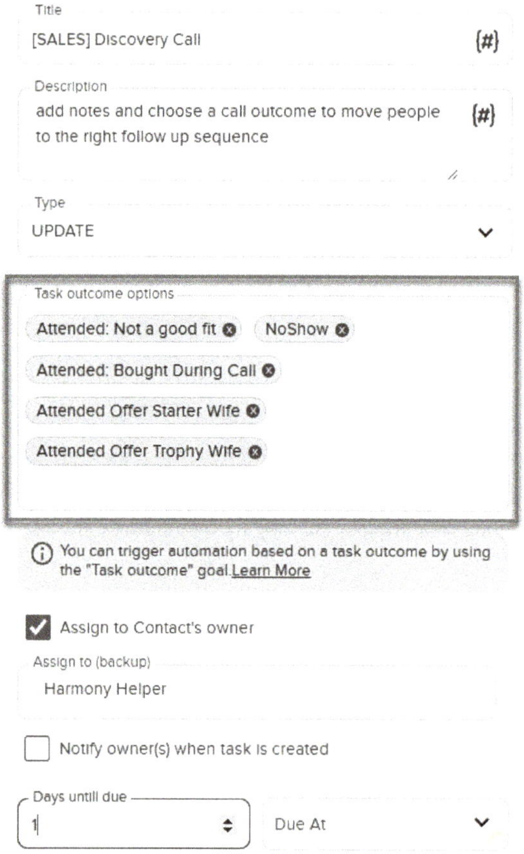

Figure 8.15 – Task setup criteria

- **Set Field Value**: This option allows you to populate a field in the contact record with a value that you've set. This is very useful when you want to set program or course start dates and can be used with your field timers to automate sending reminders or homework.

  There are four options for setting a field value:

  A. **Overwrite field value**: Overwrite existing field values with new ones.

  B. **Set field value**: Set a value in an empty field. If there is an existing date in the field, this will *not* overwrite it.

  C. **Clear field value**: Remove any data from the specified field.

  D. **Use math**: For fields that store numbers, use this to add, subtract, multiply, or divide the existing field value:

✕   **Set field value**   Save

Summary
This action allows you to automatically add information to a Contact's field, override the Contact's previous field value, or clear the existing field value.

Select an option*   ⌄

Overwrite field value
Overwrite any existing field value and replace it with a new one.

Set field value
Set a value to an empty field, this will NOT overwrite an existing field value.

Clear field value
Remove any value from the selected field (set to empty).

Use math
Use math to change an existing field value (for number fields only)

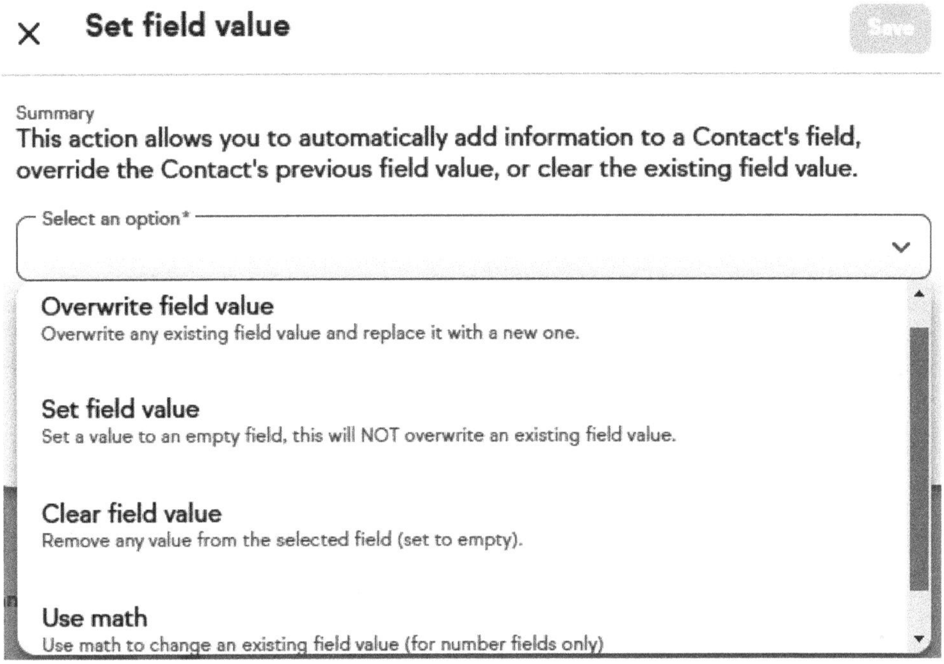

Figure 8.16 – Task setup criteria

- **Assign an Owner:** This allows you to assign a specific Keap user as the contact's owner. Doing so is an asset when you want to easily include the owners' contact details in follow-up communications.

- **Create a Deal:** The deal element allows you to add a deal to a pipeline when a lead reaches a milestone within your automation. The deal can then be tracked. As it moves through the deal stages, it can be used as an automation goal. Remember, whenever you're allowed to name something in your CRM, you want to do that. Leaving a name blank or making it generic only adds frustration later on:

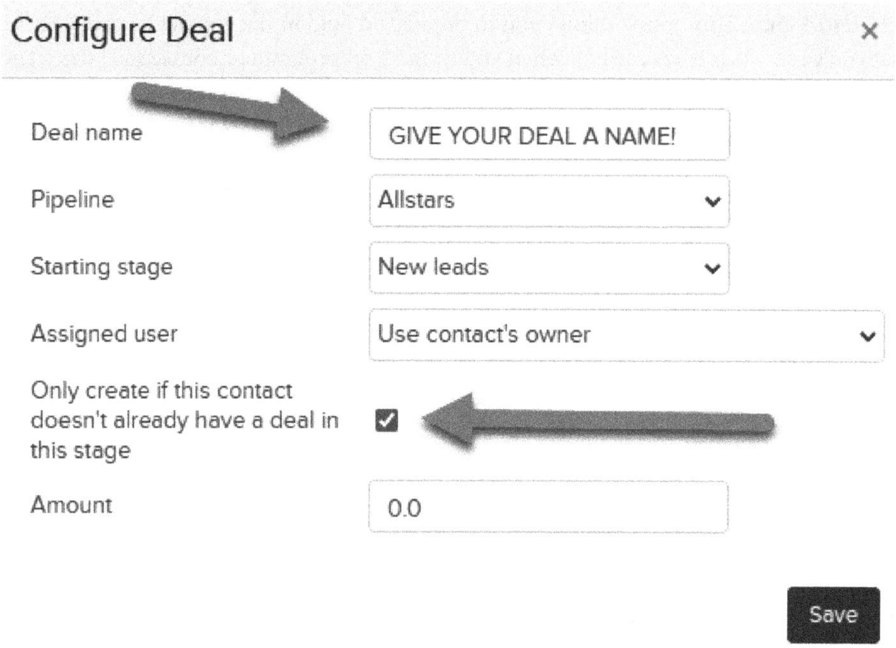

Figure 8.17 – Task setup criteria

In some cases, you may want to create a new deal even if someone has one already. Real estate is a great example of such as situation as someone may be buying or selling multiple houses. Use the checkbox to indicate that you don't want to create a deal if one already exists.

- **Create Order**: This option creates an invoice and allows you to send it as a payable invoice or send it without payment options (but seriously, why would you not want to get paid?!):

Figure 8.18 – Creating an invoice

- **HTTP Post**: This element sends information from Keap to an outside web page to trigger a script that gathers data from the URL and carries out actions on another site. An often-used example of an HTTP Post is when you want to add or update members to a membership site:

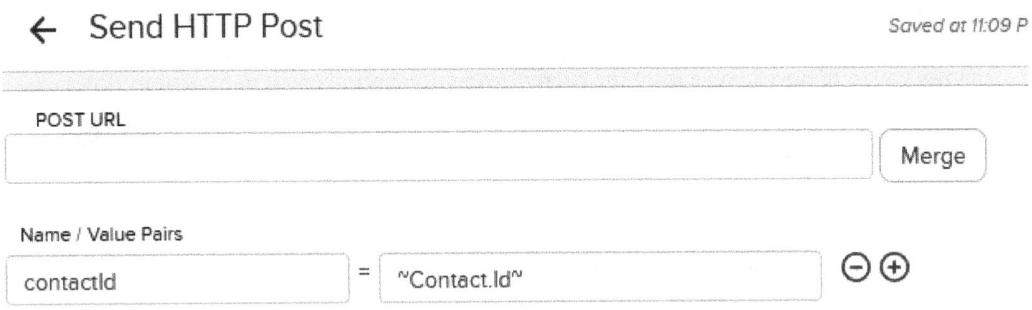

← **Send HTTP Post**   *Saved at 11:09 P*

POST URL

[   ] Merge

Name / Value Pairs

[contactId] = [~Contact.Id~]   ⊖ ⊕

Figure 8.19 – HTTP Post

- **HTTP Request**: This element takes things a step further and allows for more complex values:

    A. **Get**: Requests data from a specified resource

    B. **Post**: Sends data to a server to create or update a resource

    C. **Patch**: Applies partial modifications to a resource

    D. **Delete**: Deletes the specified resource:

← **Send HTTP Request**   *Saved at 11:19 PM*   [Test ⌄]   [Actions ⌄]   Draft

Summary
You can use Http requests to send data to any complaint server.

[Post   ⌄] [Target URL   ] Merge
| Post |
| Put |
| Patch |
| Delete|

Header

+ Add Header Parameter

Body

[   ]

Request   Response

Figure 8.20 – HTTP request features

Using HTTP Post is very useful but also requires some technical knowledge. To learn more about this feature, I recommend checking out the following resource: `https://www.w3schools.com/tags/ref_httpmethods.asp`.

- **Add to sequence**: This is quite possibly my favorite of all the elements! The **Add to sequence** element allows you to quickly add a contact to another sequence in any automation. It also allows you to remove people from the current sequence, keeping your automations tidy:

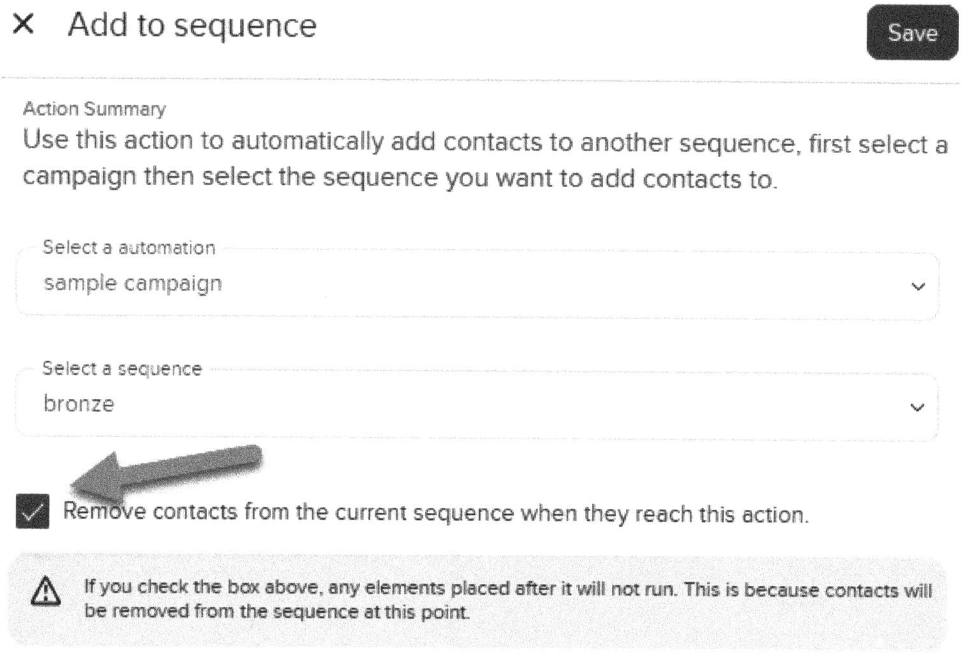

Figure 8.21 – HTTP request features

Working with decision diamonds

One of the most agile features of advanced automations is their ability to create "one-to-many" outcomes. For example, you may create an automation that invites people to purchase your membership and then offer them gold, silver, and bronze options when they view your landing page. Depending on what they purchase, you will want to tag them accordingly and send the appropriate emails so that you can follow up and deliver their membership content.

One invitation leads to many possible outcomes.

Decision diamonds are how the advanced automation builder knows which path to send your leads down. When this happens, Keap uses the rules you add to the decision diamond to evaluate the information available and use it to choose the appropriate sequence.

> **Note**
>
> Decision diamonds are created automatically when a goal or a sequence connects to two or more sequences.

How to do it...

This recipe will take you through the steps of adding rules to your decision diamond once it's been created.

Configuring decision diamonds

Using the example from the preceding recipe, let's create a one-to-many path where a purchase is made and there are three possible outcomes. We'll use three sequences to set up automation rules for each outcome. Follow these steps:

1. Start by dragging three sequences onto your canvas. Let's call those sequences **Bronze**, **Silver**, and **Gold**.

2. Next, we want to add a purchase goal to the canvas.

3. Hover over the purchase goal to grab the green arrow and drag it over your **Bronze** sequence.

4. Repeat *Step 3*, this time connecting to your **Silver** sequence.

5. A decision diamond will automatically appear between your purchase goal and your sequences.

6. Now, connect your goal and your **Gold** sequence:

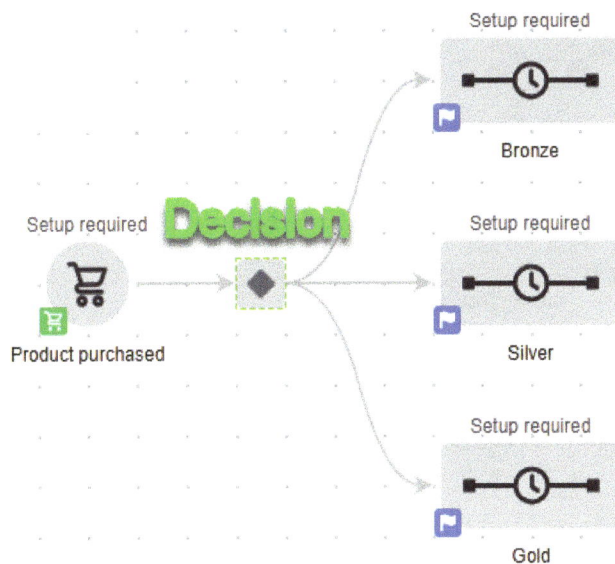

Figure 8.22 – A decision diamond is generated when a goal connects to multiple sequences

7. Double-click on the decision diamond to open it.

8. Use the + **Add a rule** button to set up the rules for each sequence. Rules can be based on tags, radio or checkbox options, custom fields, and some contact fields.

9. For this example, my rule will be **If the Contact's Tags contains bronze**. See *Figure 8.23* for clarity.

10. Repeat *Step 4* for each of your three sequences:

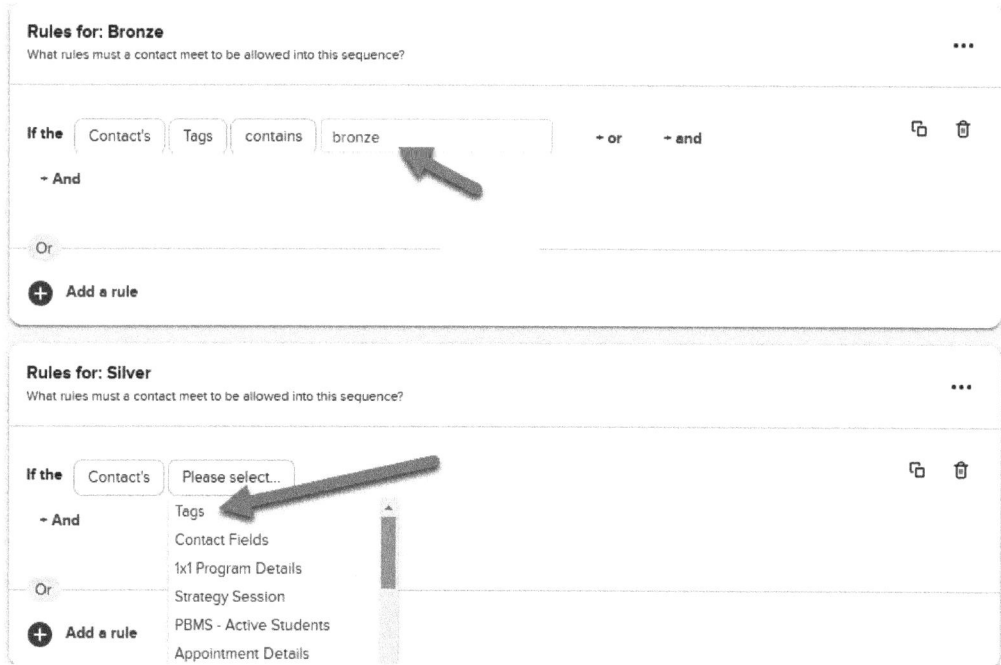

Figure 8.23 – Adding rules to your decision diamond

> **Note**
>
> If a contact matches more than one rule, they will enter multiple sequences, so it's important to consider this when creating your rules.

> **Caution!**
>
> When rules are not added to a decision diamond, it implies that everyone is included by default. Without clear criteria, it's akin to having no restrictions, granting entry to all.

This scenario, where "all rules are true," can be advantageous if you want every individual to undergo the same steps.

For example, you might want all members to receive call reminders but only want bronze members to receive a bonus and gold members to receive no bonus. In this example, you would leave the rule blank for call reminders so that everyone may enter the flow.

Take care in establishing your rules and conduct multiple tests to ensure they function appropriately.

Importing rules from another sequence

Follow these steps:

1. Double-click to open your decision diamond.

2. Open the menu via the ellipsis on the right to find a set of rules.

3. Select **Import rules from another sequence**.

4. Select another set of rules to import (you can only import rules from the decision diamond you are in):

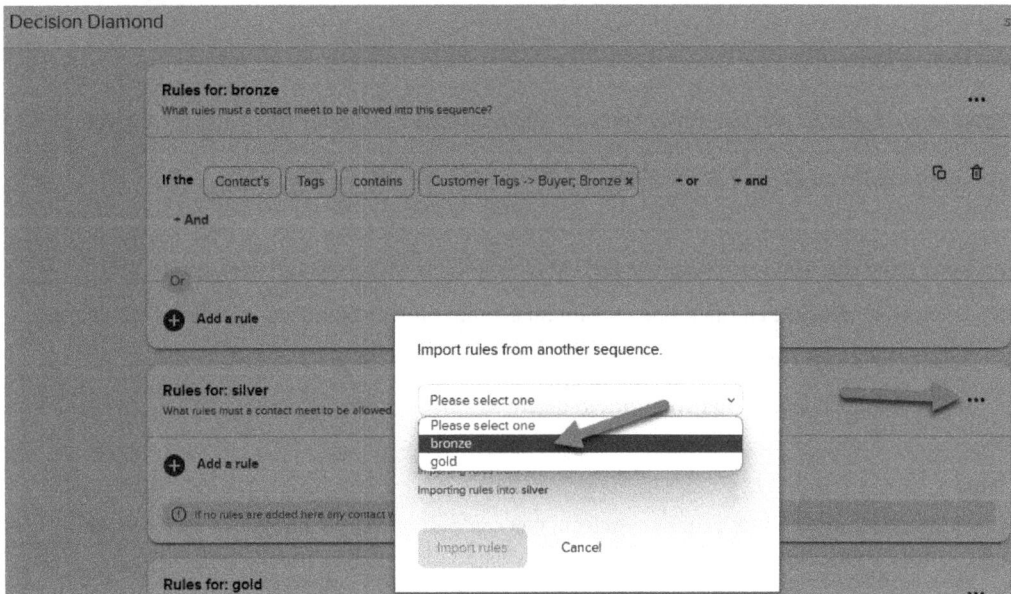

Figure 8.24 – Importing rules in a decision diamond

5. Make any necessary changes and then use the back arrow to return to your automation canvas.

Copying rules

When you need more than one rule, you can save time by copying rules rather than starting from scratch. This works best with rules based on drop-down menus. Follow these steps:

1. Double-click to open your decision diamond.

2. Click the **copy** icon next to an existing rule.

3. Modify the new rule as needed:

Rules for: bronze

What rules must a contact meet to be allowed into this sequence?

Copy rule

If the Contact's Tags contains Customer Tags -> Buyer: Bronze **x** ᐩ **or** ᐩ **and**

ᐩ **And**

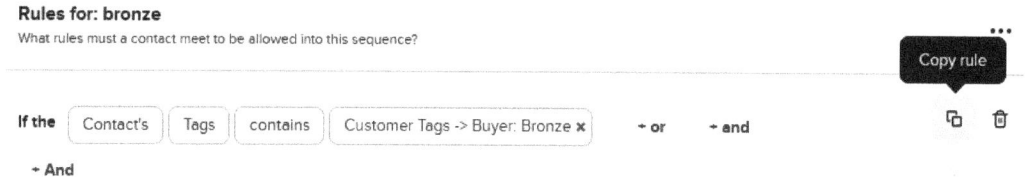

Figure 8.25 – Importing rules in a decision diamond

Deleting rules

Sometimes, it's easier to just start over. You can delete individual rules by following these steps:

1. Open the menu to see a set of rules.

2. Double-click to open your decision diamond.

3. Locate the rule you want to delete and then click the trash can icon to remove it.

4. If you wish to delete all rules, open the menu for the rule by clicking on the ellipsis on the right.

5. Select **Delete all rules**.

6. In the popup, confirm that you want to delete all existing rules.

> **Note**
>
> If you want to completely wipe the slate clean, you can disconnect the lines from each sequence and rebuild your connections to spawn a new decision diamond.

How it works...

With this recipe, you've got all the tools to unlock the true power of Keap's advanced automation features. Now that you're familiar with conditional logic, multi-step workflows, dynamic content, and multi-channel integration, you're ready to craft personalized and impactful automation sequences that are perfectly aligned with your business goals.

With the addition of robust tracking and reporting features, you can easily evaluate and optimize your automation strategies for maximum efficiency and results.

Building an advanced automation

Now that we have a good grasp on the tools needed, let's build our first advanced automation!

How to do it...

Let's begin with a form offering a free ebook titled, *3 reasons why cats rule*. In exchange for the user providing their first name and email address, we will send them our free ebook:

1. Click on the **AUTOMATION** tab in the left-hand side navigation bar to open the menu and choose **AUTOMATION BUILDER**.
2. Click the + sign to start a new automation.
3. In the popup that appears, give your automation a name.
4. Select a category for your automation or create a new one.
5. Click **Save**.
6. Click on **Automations** in the left-hand menu. Click the + button to create a new automation and choose **Advanced Automation**.
7. Drag a **Form submitted** goal onto the canvas.
8. Double-click the **Form submitted** goal to customize your form or choose one you've already created. By default, all forms have first name and email already built into them.
9. After customizing your form, click the back button to return to your automation.

Now that we have a web form goal, it's time to connect it to an automation sequence:

1. Drag a sequence onto the canvas to the right of the web form.
2. Grab the green arrow on the web form goal and drag it to the sequence to connect.
3. Rename the sequence `Welcome and deliver`.
4. Double-click the sequence to open it.
5. You are now viewing the sequence canvas. A **Start** tool has already been added to the sequence.
6. Drag an **Apply/Remove Tags** element onto the canvas. Double-click the tag to open it. We're going to add a tag that signifies that the contact opted in for our ebook.
7. Click inside the **select a tag** box. You can search for a tag by typing in its name:

    A. For this recipe, we'll create a tag called `submitted: cat ebook`.
    B. If the tag exists, you will see it in the list. Click to add it to the box.
    C. If the tag doesn't exist, you will see a + sign. Click it to create the tag and save it:

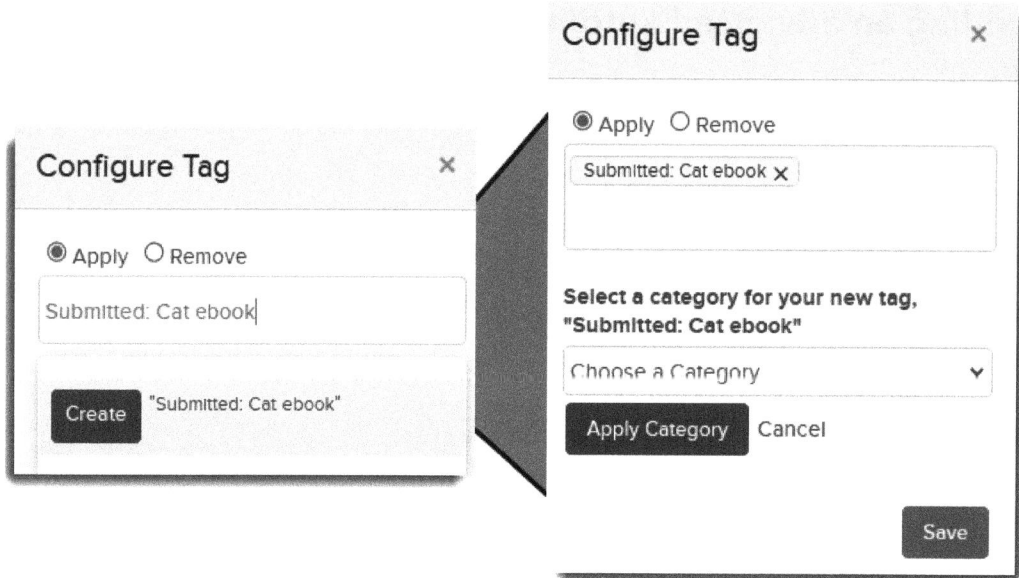

Figure 8.26 – Adding a tag within the advanced automation builder

8. Next, drag an email element onto the canvas to the right of the tag. It will automatically attach to the previous element.

9. Double-click where it says **untitled email** and give your email a name. I like to use the subject of the email as my name, so this email will be titled **welcome – ebook link inside**.

10. This email will be sent immediately when someone submits the web form (accomplishes the goal).

11. Double-click to open the email builder and follow the instructions provided in *Chapter 4, Communicating with Your Lists*.

12. Once you're satisfied that your email is complete, move your email from draft to ready and close the email builder to continue.

13. Knowing that people sometimes don't open their emails, let's plan on sending a reminder. Drag a delay timer onto the canvas to the right of our email.

14. Double-click to open the timer.

15. Set the timer to wait **1 Day(s)** and send **Any Day** at **8:00 AM**:

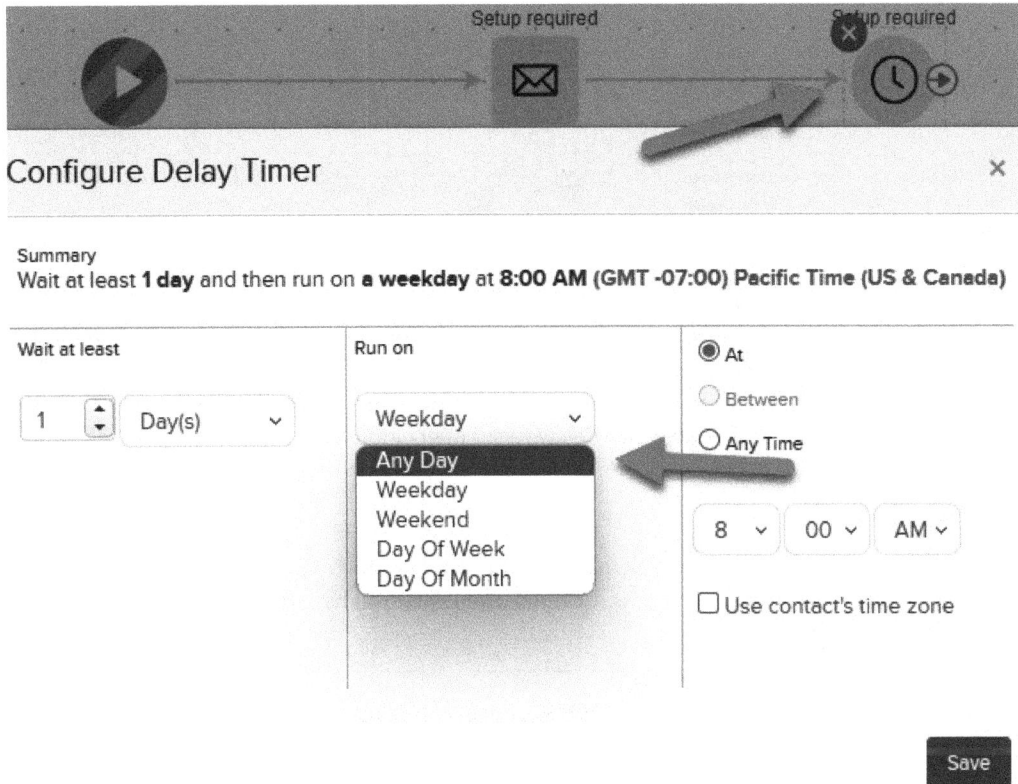

Figure 8.27 – Configuring the delay timer

16. Click **Save** to continue.

17. Right-click on our previously added emails and choose **Duplicate**.

18. Click and drag the copy to the right of our delay timer.

19. Grab the green arrow on the delay timer and drag it to the email to connect it.

20. Open the email and edit the subject line to add the word `reminder` to it.

21. Click and drag a **Delay Timer** element onto the canvas to the right of our email.

22. Double-click the **Delay Timer** icon and set it to run 5 days later.

23. Click **Save** to return to your sequence.

24. When you're done configuring the element in your sequence, be sure to toggle them to **ready** where applicable:

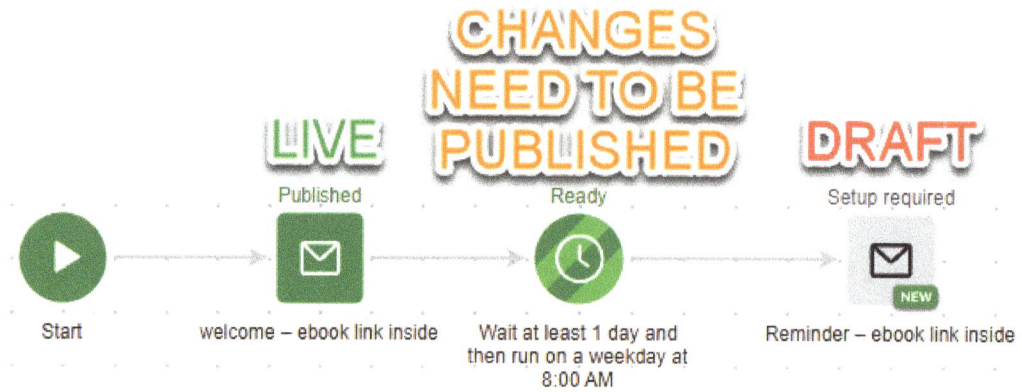

Figure 8.28 – Ready versus draft mode

> **Note**
>
> Keap makes it easy to know if your elements are in ready mode or not with the use of color blocking. Solid green means it's live, green striped means that you have unpublished edits, and gray means it's in draft mode.

25. Toggle your sequence to **ready** and exit the sequence to return to your automation canvas.

26. Publish your automation.

But we're not done yet! We've completed the lead capture and list-building part but the ultimate goal of our funnel (or any funnel) is to convert them into paying clients:

1. So, let's add another sequence after the **Welcome and deliver** step and call this one **Offer air biscuits fan club**.

2. In this sequence, we're going to add three to five emails with a 2-day delay timer between each email:

Figure 8.29 – Offer air biscuits fan club

Remember, we want to entice people to buy our product, but we also want to stop sending them emails once they make a purchase.

3. Drag a purchase goal onto the automation canvas and link it to the offer sequence. Let's call it **air biscuits fan club**:

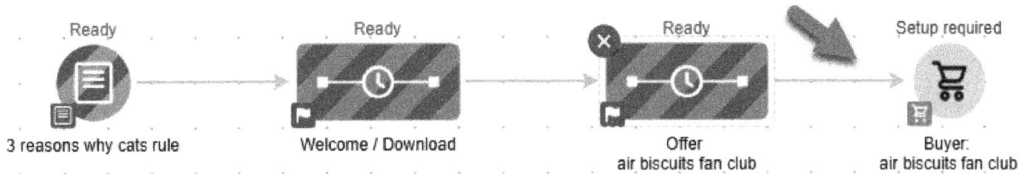

Figure 8.30 – Basic campaign setup

4. Double-click the element to set up the purchase goal.

5. In this recipe, the goal is satisfied when someone purchases a specific product:

Figure 8.31 – Setting purchase criteria

6. Add the types of payments you will accept.

7. Click **Save**.

> **Stop**
>
> The next step is to publish your automation. However, before you do, it's a very good idea to take a minute and look at all your automation pieces and make sure you've given everything a proper name that is consistent with your naming convention. Good naming habits save you from the frustration of trying to find things later on only to realize you have dozens of forms, emails, and deals named "untitled."

8. Publish your automation by clicking the **Publish** button on the top right of the automation canvas.

9. The automation builder will validate your campaign and let you know if there is anything that needs to be fixed before it can be published:

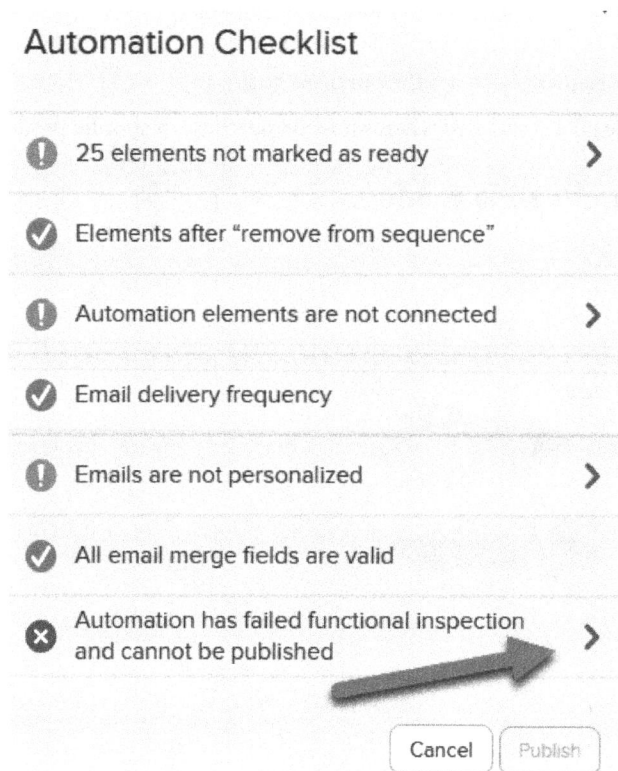

Figure 8.32 – Automation errors can't be published

Let's take a closer look:

A. Green icons indicate automation elements that are good to go

B. Yellow icons warn that you might want to review the steps but they will still publish

C. Red icons indicate parts of our automation that cannot be published until they are fixed.

10. The item in red cannot be published. Click the > icon next to the red item to see what's broken.

11. Click the eyeball icon to quickly access the broken item:

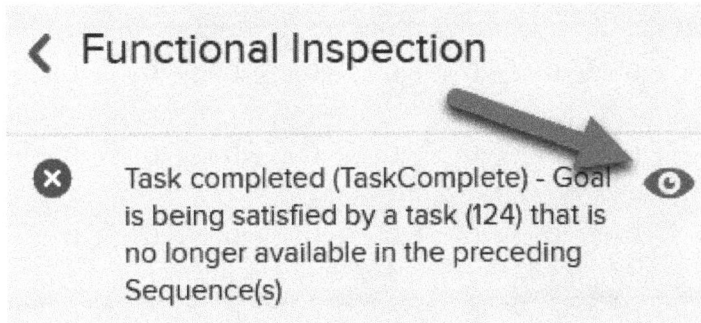

Figure 8.33 – Automation errors can't be published

12. After correcting any mistakes, click **Publish** again to go live with your automation.

How it works...

In Keap, creating advanced automations is an exciting adventure where you have the power to design and implement every aspect of your sales funnel. Begin by establishing a lead capture goal and connecting it to a sequence, where you can deliver engaging follow-ups and nurturing content. Then, add a touch of automation magic by defining conversion goals, such as product purchases. Before unleashing your automation to transform your marketing and sales endeavors, ensure everything is polished and primed to captivate your audience.

9

Reports

In a CRM, reports are like your secret weapon, giving you the inside scoop on different parts of your business, such as how sales are going, how engaged your customers are, and how effective your marketing is. When you dive into the numbers and trends in these reports, you can make smart decisions, spot where you can do better, and fine-tune your strategies to make your business thrive.

In this chapter, we'll learn how to utilize the reports that Keap provides by covering the following recipes:

- Sales reports
- Contact tracker reports

Technical requirements

For this chapter, the following skills may be helpful:

- **Basic Excel skills**: CRM reports operate similarly to Excel, so skills such as sorting, filtering, and basic formulas can be very useful
- **Attention to detail**: Paying attention to detail is crucial for accurately interpreting CRM reports and avoiding errors in data entry or analysis

Let's get ready to use reports in Keap! CRM provides a series of essential steps to ensure you're getting the most out of your data.

First, it's all about getting cozy with the platform itself. Throughout this book, we've been getting to know the ins and outs of Keap; from entering data to creating a pipeline and mastering navigation. Everything has been building up to this moment – that is, mastering the app to the point where your skills are sharp enough to navigate reports and truly understand your business more deeply.

Now, it's time for you to fully understand your reporting objectives and understand the nitty-gritty of your data structure to ensure your reports align perfectly with your business goals and maintain data integrity.

Working with sales reports

In a CRM, sales reports are your compass, offering a full view of your pipeline. They provide metrics such as leads, deals, and revenue over a specific period. These reports show how your sales teams and reps are performing and give insight into the overall sales process. They're essential for tracking progress, spotting trends, predicting future sales, and making data-driven decisions to improve efficiency and effectiveness.

How to do it...

Financial or sales reports are a quick way to see what's happening in your business. You can also check out your sales data graphically using the **All sales report** widget in Keap.

Using the All sales report widget

1. Click on the **Reports** tab in the left-hand side navigation bar.

2. Click the **All sales report** link on the widget:

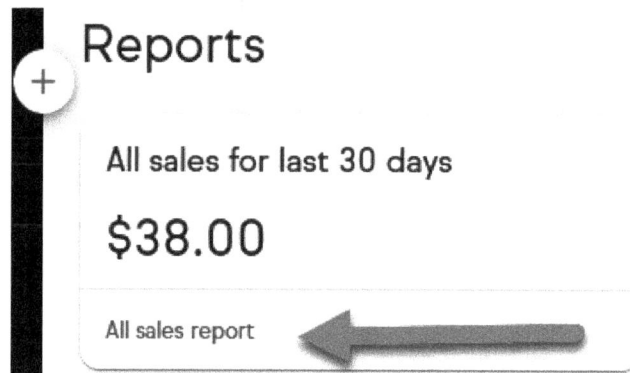

Figure 9.1 – The All sales report widget

3. The **All sales report** widget will display the following sales totals:

    A. **Total sales**

    B. **Average order value**

    C. **Total number of sales**

4. Use the **Showing data for** dropdown to select the time period you want to analyze:

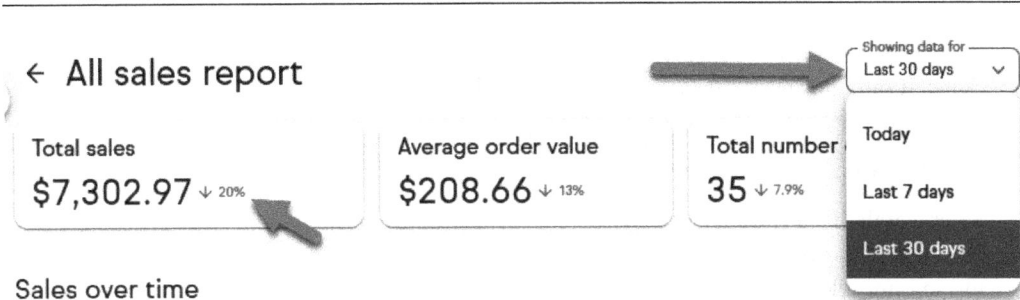

Figure 9.2 – All sales report comparison

5. Report values will be updated to reflect the date range you choose and a comparison to the previous period will be displayed next to the dollar amount.

A graph will also appear below the widget to provide a visual representation of sales values:

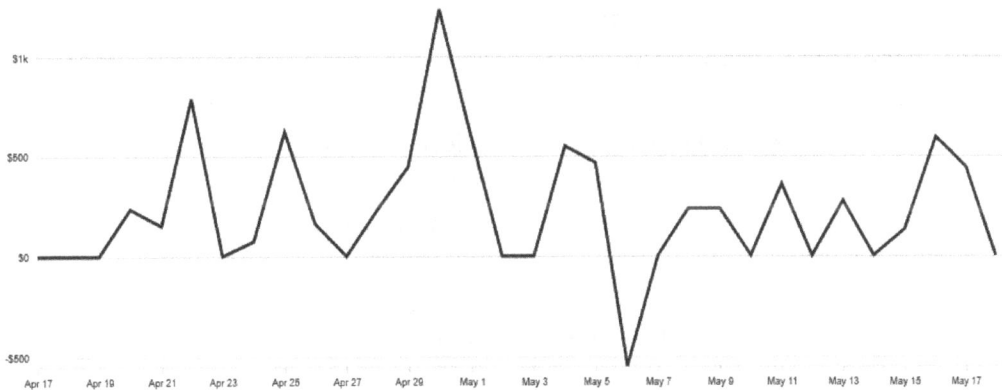

Figure 9.3 – All sales report graph

Below the graph, you will see a list of all orders within the selected timeframe.

Generating a sales report

Keap offers six prebuilt sales reports, each providing a distinct perspective on your sales and receivables. Although these reports might appear similar at first glance, each serves a unique purpose and thus comes with its own set of filters and customization criteria:

*   **All Sales Report**: This report returns the data on all sales but doesn't display the results grouped by type, month, year, and so on like **Payments Report** does.

*   **All Sales (Itemized) Report**: This report gives a more detailed breakdown, including product information, shipping, discounts, refunds, and more.

- **Payments Report**: This report shows all payments that were made and can be grouped by date range. It can find specific transactions based on the criteria you choose.

- **Receivables**: This report shows a detailed list of contacts with a balance due. It can help you keep an eye on your cash flow.

- **Sales Totals (By Product)**: This report is great for figuring out which are your top-selling products. It shows the number of orders per product, the total count of products ordered, and the amount sold grouped by product.

- **Failed Invoice Report**: This report displays all failed auto-charge transactions for subscription programs and payment plans.

Viewing a sales report

Follow these steps:

1. Begin by navigating to the **Reports** section by clicking the **Reports** icon on the navigation panel.

2. Click anywhere in the **View all reports** box.

3. Search for `sales` to filter the options.

4. Click the arrow to the right of the report name you want to view. For this recipe, I will be using **All Sales (Itemized) Report** to show any sales that have an open balance. The workflow presented here is the same for any report you wish to pull.

5. Reports in Keap open to the last search values used. To continue, click **New Search** to clear any previous report data:

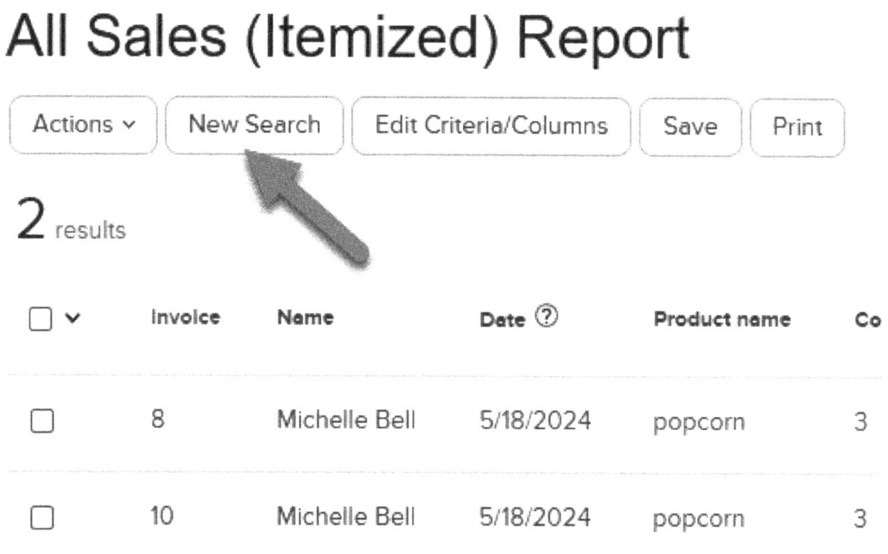

All Sales (Itemized) Report

| Actions ∨ | New Search | Edit Criteria/Columns | Save | Print |

2 results

	Invoice	Name	Date ⑦	Product name	Co
☐	8	Michelle Bell	5/18/2024	popcorn	3
☐	10	Michelle Bell	5/18/2024	popcorn	3

Figure 9.4 – Starting a new report search

6. The **All Sales (Itemized) Report** option offers many options for search criteria. They are grouped into four tabs:

    A. **Search**

    B. **Address**

    C. **Misc Criteria**

    D. **Custom Fields**

7. Use the provided filter options to input your search criteria to narrow your results. Since this example involves pulling a report for orders with a balance, you can click on the **Misc Criteria** tab and add criteria so that **Balance** must be **greater than** $5:

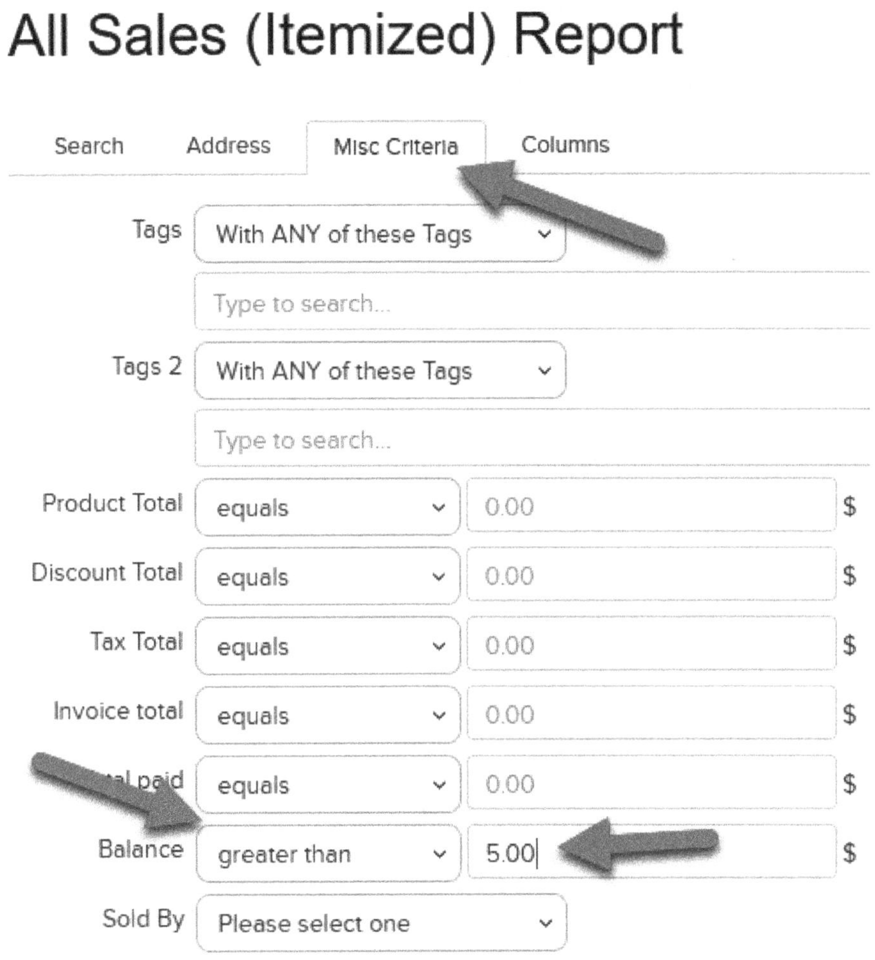

All Sales (Itemized) Report

Search	Address	Misc Criteria	Columns

Tags	With ANY of these Tags ⌄	
	Type to search...	
Tags 2	With ANY of these Tags ⌄	
	Type to search...	
Product Total	equals ⌄	0.00	$
Discount Total	equals ⌄	0.00	$
Tax Total	equals ⌄	0.00	$
Invoice total	equals ⌄	0.00	$
al paid	equals ⌄	0.00	$
Balance	greater than ⌄	5.00	$
Sold By	Please select one ⌄	

Figure 9.5 – Adding report criteria

8. The **Columns** tab enables you to customize your report view by adding or removing data columns from the results. Here are the steps to do so:

    I.   Click the **Columns** tab.

    II.   Click the +**Add a field** link.

    III.   In the search box, type `balance` to filter the available criteria.

    IV.   Click on the word **Balance** to add the field to your report.

    V.   You can click the minus symbol to the right of any criteria to remove it from your report.

    VI.   You can also click and drag the dots to the left of each field to re-order the report to your liking·

Customize Columns

This allows you to edit the presentation of the search results. The fields below represent the current columns that will be displayed in your search results. To add more columns, select the "Add a field" at the bottom of the list. To re-order the columns, simply drag and drop.

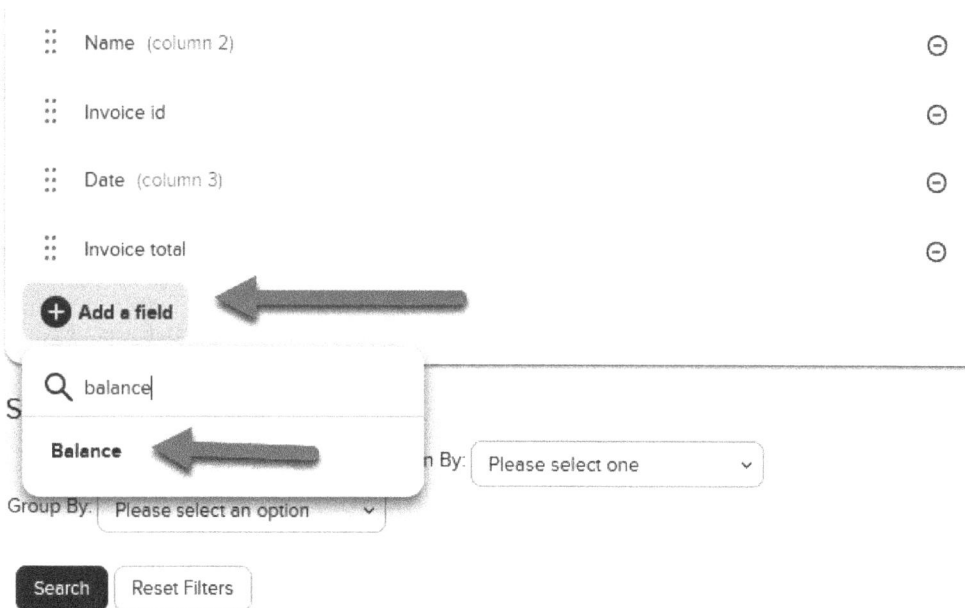

Figure 9.6 – Adding report criteria

9. Once you've selected all your criteria, click **Search** to view your report:

> **Note**
> Using the **Reset Filters** button will clear your criteria so that you can restart your search.

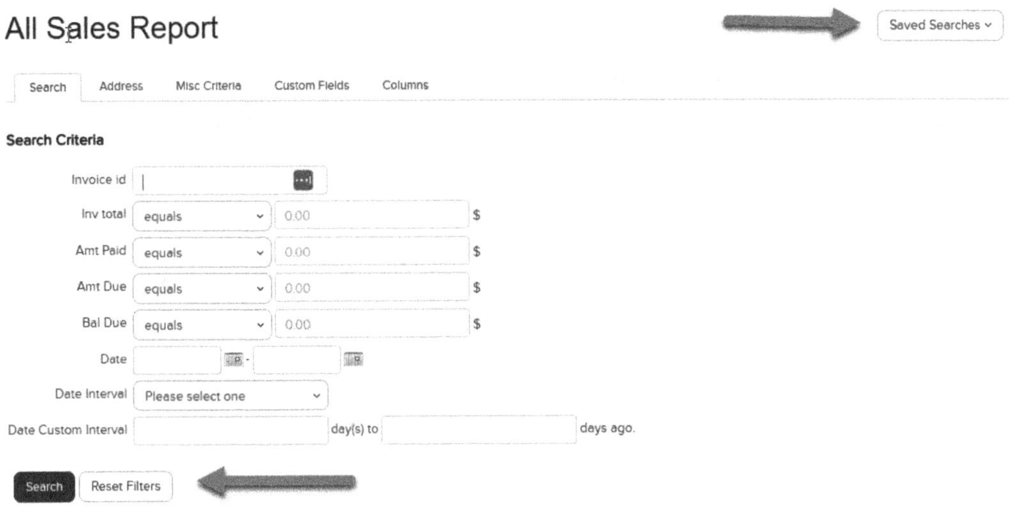

All Sales Report

Figure 9.7 – All Sales Report

Saving your sales report search criteria offers several benefits beyond just time-saving. By preserving your search settings, you ensure consistency in data analysis, allowing for accurate comparisons and trend tracking over time. Additionally, saved searches enable quick access to valuable insights, facilitating informed decision-making and enhancing overall productivity in managing sales performance.

Saving a sales report

Follow these steps:

1. Click the **Save** button.

2. A pop-up box will open:

    I.   Give your report a short descriptive name. This name is what you will see in the saved searches menu.

    II.   Saved reports can be shared with other Keap users in your app. In the **Who would you like to share this report with** box, you can do the following:

    •   Select **everyone**

    •   Hold down the *Shift* key to select a group of users

    •   Hold down the *CMD* key on a Mac or the *Ctrl* key on a PC to select individual names from the list

    III. Click **Save**.

Viewing a saved sales report

1. Click on the **Reports** tab in the left-hand side navigation bar.

2. Locate the report type and click the arrow to the right of the report's name.

3. Click the **Saved Searches** button to open the dropdown

4. Type to search or scroll to find your search:

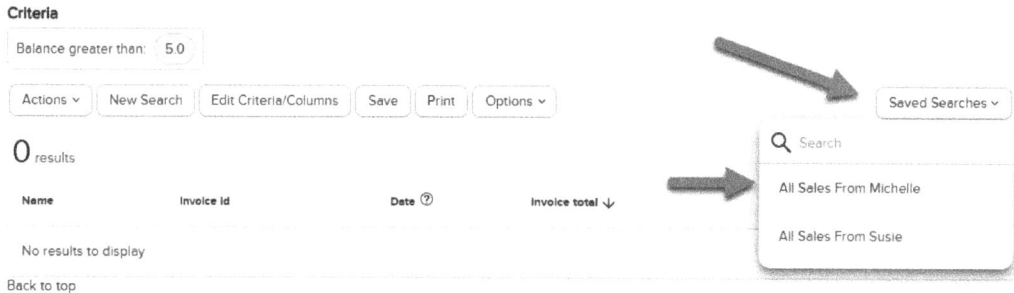

Figure 9.8 – Saved searches and reports

How it works...

Regularly reviewing your sales reports in Keap provides you with valuable insights into your sales performance, helping you track progress, identify trends, and make informed decisions. These reports enable better sales management, facilitate forecasting, and ultimately contribute to driving growth and success.

Working with contact tracker reports

In Keap, we handle contact tracking through reports by bringing together contact details and interactions, giving users the ability to delve into data such as contact engagement, conversion rates of leads, and customer demographics. Through these reports, users gain valuable insights into contact behavior, sales potential, and the effectiveness of marketing efforts. This empowerment allows businesses to foster connections, pinpoint avenues for growth, and customize strategies to meet the needs of their contacts more effectively.

How to do it...

Keap automatically tracks various interactions with contacts, including email opens, link clicks, form submissions, and more. Contact tracking reports are your window into knowing how your contacts are engaging with your business, providing valuable insights into their behaviors, preferences, and levels of engagement.

Generating a contact tracker report

Follow these steps:

1. Click on the **Reports** tab in the left-hand side navigation bar.

2. Under the **contact tracker** header, you will find the following reports:

    A. **Tag Tracker**: This report shows you when tags were applied to a specific contact

    B. **Web Form Tracker**: This report shows you when specific contacts submitted forms

    C. **Web Form Engagement Tracker**: This report summarizes new, unique, and repeat contact activity for each web form

    D. **Sent Email Tracker**: This report provides a historical list of sent emails with engagement stats for each contact

    E. **Email Batch Results**: This report displays information about all sent emails, including bounced emails

    F. **Email Engagement Tracker**: This report allows you to view and manage contact email engagement stats and confirmation status

    G. **Unsubscribe Tracker**: With this report, you can see who has unsubscribed from your lists

    H. **Campaign Enrollment Tracker**: This report displays unique contacts who are enrolled in specific automation

    I. **Campaign Engagement Tracker**: This report shows you who is engaged in active automation and currently receiving content from you

    J. **Campaign Progression Tracker**: With this report, you can see which contacts are waiting for a step in an automation sequence

    K. **Campaign Sequence Recipients**: This report shows what contacts have received an email or text within automation sequences

    L. **Campaign Goal Tracker**: With this report, you can see which of your contacts have completed a goal in automation

Viewing a contact tracker report

Follow these steps:

1. Click the arrow to the right of the report name you want to view. For this recipe, let's choose **Tag Tracker**.

2. Reports in Keap open to the last search values used. To continue, click **New Search** to clear any previous report data.

3. The **Tag Tracker** report has fewer options for search criteria compared to the sales reports, so you will only see the **Search** tab.

4. Use the provided filter options to input your search criteria to narrow your results. Let's use the **Date Applied** dropdown and choose **Last 7 Days**:

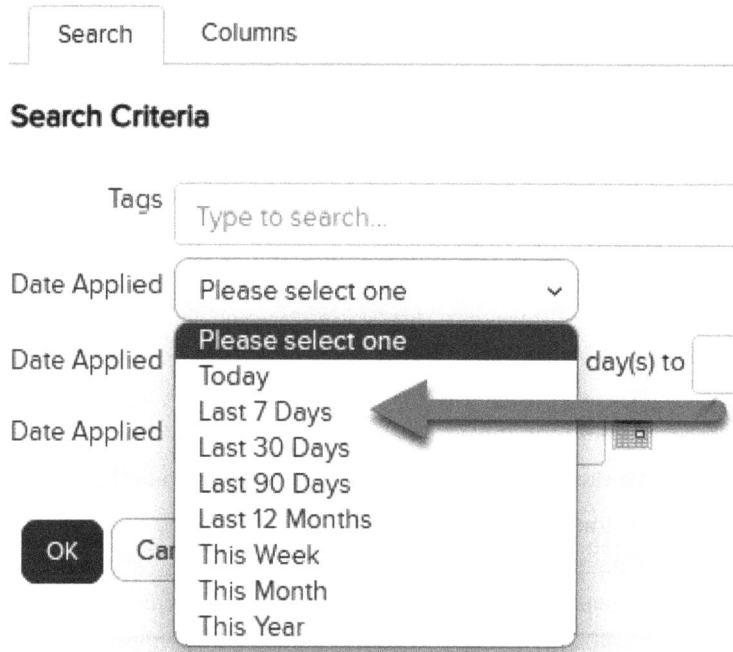

Figure 9.9 – Adding search criteria

5. Once you've selected all your criteria, click **OK** to view your report.

Saving a contact tracker report

Follow these steps:

1. Click the **Save** button.

2. A pop-up box will open:

    I.   Give your report a short descriptive name. This name is what you will see in the saved searches menu.

    II. In the sharing box, you can choose to share with everyone, select individuals, or only yourself.

    III. Click **Save**.

Viewing a saved contact tracker report

Follow these steps:

1. Click on the **Reports** tab in the left-hand side navigation bar.

2. Locate the report type and click the arrow to the right of the report's name.

3. Click the **Saved Searches** button to open the dropdown.

4. Type to search or scroll to find your search.

How it works...

Keap's suite of insightful reports simplifies contact tracking, gathering vital data on contact engagement, lead conversions, and customer demographics. When you explore reports such as **Tag Tracker** and **Web Form Engagement Tracker**, you can uncover valuable insights about your leads and clients.

Whether you're nurturing relationships or maximizing growth opportunities, mastering the agility of setting filters and accessing various data is crucial in the world of reporting. With Keap's customization options, you're empowered to generate tailored insights, guide strategic decisions, and ignite success for your business.

Understanding how contacts in your CRM interact with your content, engage with your emails, and progress through your funnels. The clue to pivot is when something isn't working, at which point you can amplify your efforts when they always reside within your reporting data!

Part 5: Integration and Optimization

In this final part, you'll explore strategies for integrating Keap with other tools and systems, ensuring seamless operations and maximizing its potential across your business.

- *Chapter 10, Tying it All Together*
- *Chapter 11, Five Essential Automation Funnels*
- *Chapter 12, Data Management and Maintenance*

10

Tying It All Together

Now that your Keap account has been set up, let's take a peek into the life of a Keap user – a day brimming with efficiency and productivity. Kickstart your morning by logging in to the platform and glancing at your dashboard, where you will catch a glimpse of key metrics, upcoming tasks, and recent customer activity.

As the day unfolds, you will seamlessly transition between managing contacts, automating workflows, and crafting targeted email campaigns. Harnessing the full power of Keap's features, you will nurture leads, track sales opportunities, and interact with customers – all in a day's work.

In this chapter, we will dive into applying the most commonly used functions to your daily routine. We will cover the following recipes:

- Uploading files
- Recording payments
- Manually adding and/or removing contacts from an automation
- Sending invoices and quotes
- Adding notes to a contact record
- Adding tasks to a contact record
- Creating a deal
- Requesting Google Reviews
- Customizing your dashboard

Technical requirements

For this chapter, the following skills may be helpful:

- Depending on your business needs, you may require integrations with other software or services such as Zapier, Stripe, or QuickBooks.

- In *Chapter 2*, we went over how to link Google Reviews to your Keap dashboard. It's important to complete this step before moving forward because we'll be diving into how to request reviews.

Before we get into the *day in the life* of a Keap user, let's ensure you're ready to roll. If you haven't already, take a moment to familiarize yourself with Keap's essential features. This will make the transition from implementation to daily usage smoother. Our goal is to create your optimal workspace, encompassing everything from the dashboard layout to easily utilizing essential functionalities.

Managing documents and files

With Keap, you can easily upload files directly to a contact record, ensuring you never misplace vital information about your leads and clients. Whether it's documents, images, or anything else under 10 MB, you can securely store and access them anytime from the contact record. It is a seamless way to keep everything organized and at your fingertips!

How to do it...

Keap supports several file types, making data retention and delivery simple. Familiarizing yourself with what types of files you can upload will save you time and effort later on. There's no sense in trying to upload files that cannot be uploaded.

Here's a list of all the file types:

Supported File Types							
Data	Image		Audio	Video		Text	
.csv	.gif	.tif	Audio	.mp4	.mp4v	.doc	.notes
.xlr	.jpeg	.tiff	.wav	.mpg	.wmv	.docx	.log
.xls	.jpg	.jif	.wma	.mpeg	.avi	.pages	.msg
.xlsx	.png	.jiff	.mp3	.mov	.m4v	.rtf	.xdl
.key	.pdf	.jp2	.mp4	.movie	.mpeg1	.wpd	.wp
.pps	.psd	.jpx	.mid	.mpeg4	.wm	.wps	.wp4
.ppt	.fpx	.j2k		.flv	.f4p	.err	.wp5
.pptx	.pcd	.thm		.rm	.hdv	.txt	.wp7
.xml	.bmp	.yuv		.vob	.divx	.text	.wsd
	.pspimage			.qt	.gvi	.pwd	
				.hdmov	.m2ts		
				.mnv			

Figure 10.1 – Supported file types

In this recipe, we will cover managing documents and files within your contact records.

Uploading files to a contact record

Follow these steps to learn how to upload files to a contact record:

1. Using the navigation or search feature, locate the contact record you want to upload a file to.
2. Click the **More** icon to open the drop-down menu.
3. Select **Upload a file**.
4. Select the file you want to upload:

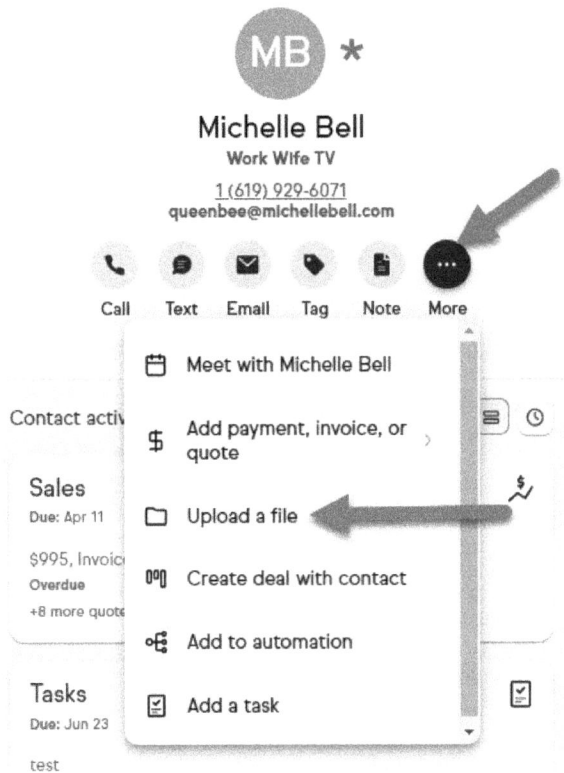

Figure 10.2 – Uploading a file

Viewing uploaded files

Follow these steps:

1. From a contact record, locate the **Files** card under **Contact activity**, as shown in *Figure 10.3*.
2. Clicking on the card will open a side menu.

3. Use the search box to locate your file if needed.

4. You can now download the file to view it.

Renaming an uploaded file

Follow these steps:

1. Clicking the ... (ellipsis) icon to the right of the download icon opens a drop-down box.

2. Select **Rename file**.

3. In the provided pop-up box, edit the name of your file.

4. Click **Save**:

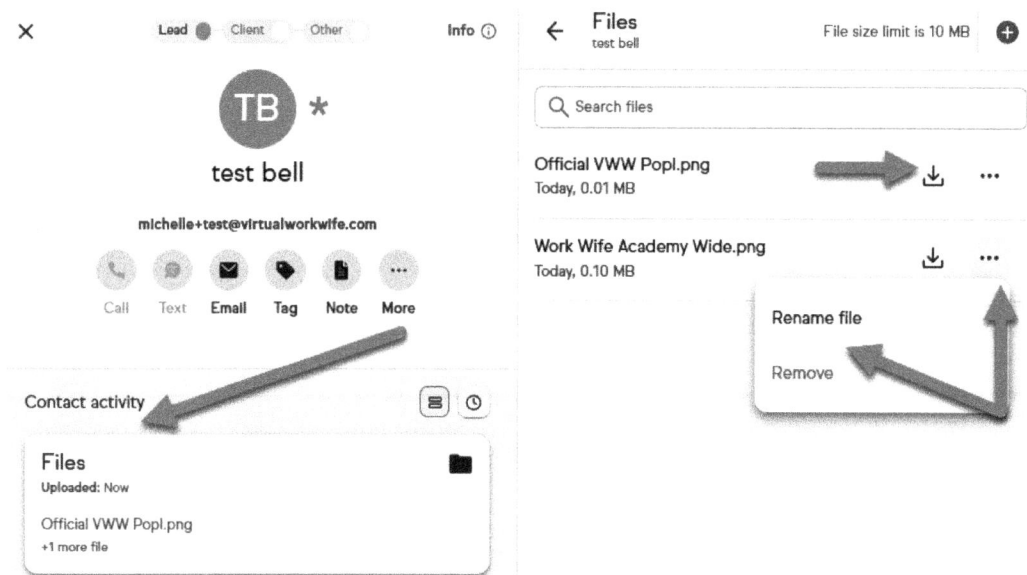

Figure 10.3 – Working with uploaded files

Deleting an uploaded file

Follow these steps:

1. Clicking the ... (ellipsis) icon to the right of the download icon opens a drop-down box.

2. Select **Remove**, as shown in *Figure 10.3*.

3. In the provided pop-up box, confirm this action by selecting **Delete**:

Figure 10.4 – Deleting uploaded files

Sending an uploaded file to a contact

Follow these steps:

1. Using the navigation or search feature, locate the contact record you want to upload a file to.

2. Click the **Email** icon.

3. In the email editor popup, provide details in the **To** and **Subject** areas and add a message. You can use the **Signature** toggle to add your signature if desired.

4. Click the paperclip icon in the lower left corner to upload your document.

5. Click **Save** to send your email with the attached file.

6. Keap will automatically add your document to the **Files** card in the contacts activity, as shown in *Figure 10.3*:

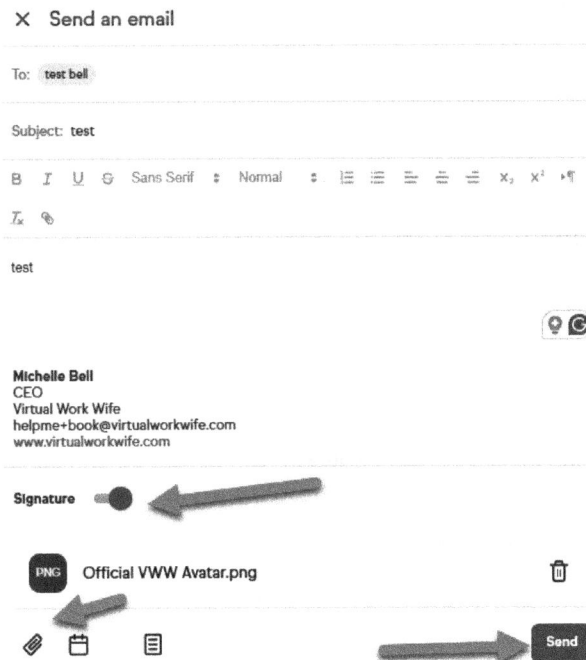

Figure 10.5 – Sending a document via email

How it works...

Keap makes uploading files to a contact record a breeze. Whether it's an essential document or a handy resource, simply upload files under 10 MB and, rest assured, you'll never lose track of vital information again. Need to access it later? Keap's got you covered – simply download it from the contact record whenever you need it.

You can also craft tailored emails directly from a contact record. You can add attachments, infuse them with your signature, and elevate them with the rich text editor to make them pop. The cherry on top is that you have the flexibility to send emails from any email address linked to your Keap user accounts. And the best part? The From: address mirrors the user record you select in the drop-down menu. It's simple, seamless, and oh-so-efficient.

Posting manual payments

While Keap does a stellar job of snagging payments automatically through invoices and checkout forms, there are times when you still need to log a manual payment. Picture this: your neighbor swings by with cash in hand, or a client spills their credit card digits over the phone. You've got to jot down those payments pronto to keep your records shipshape and nudge folks along your automation journey.

Once that payment hits, Keap swoops in and does the heavy lifting. It automatically updates your invoice status to "paid" and neatly logs all the payment details for your reference. It's like having a trusty assistant on standby, making sure your financial records stay on point while you focus on growing your biz.

How to do it...

In this recipe, we will cover how to manage manual payments.

Adding a manual payment

Follow these steps:

1. Using the navigation or search feature, locate the contact record you want to record a payment for.

2. You have two options for accessing payments:

    A. If the contact has made purchases, you will see a **Sales** card in their activity feed. Click to open the **Sales** sidebar and then click the **Add a payment** button.

    B. Click the **More** icon to open the dropdown and choose **Add payment**, **Invoice**, or **Quote**.

    Either option will open the payments pop-up menu.

3. Click on the **Payment reference** box. You may now do one of the following:

    A.   Click **Select an invoice** to choose a pre-existing invoice.

    B.   Click **Select a product** to create an instant invoice.

    For this recipe, we are going to choose **Select an invoice**:

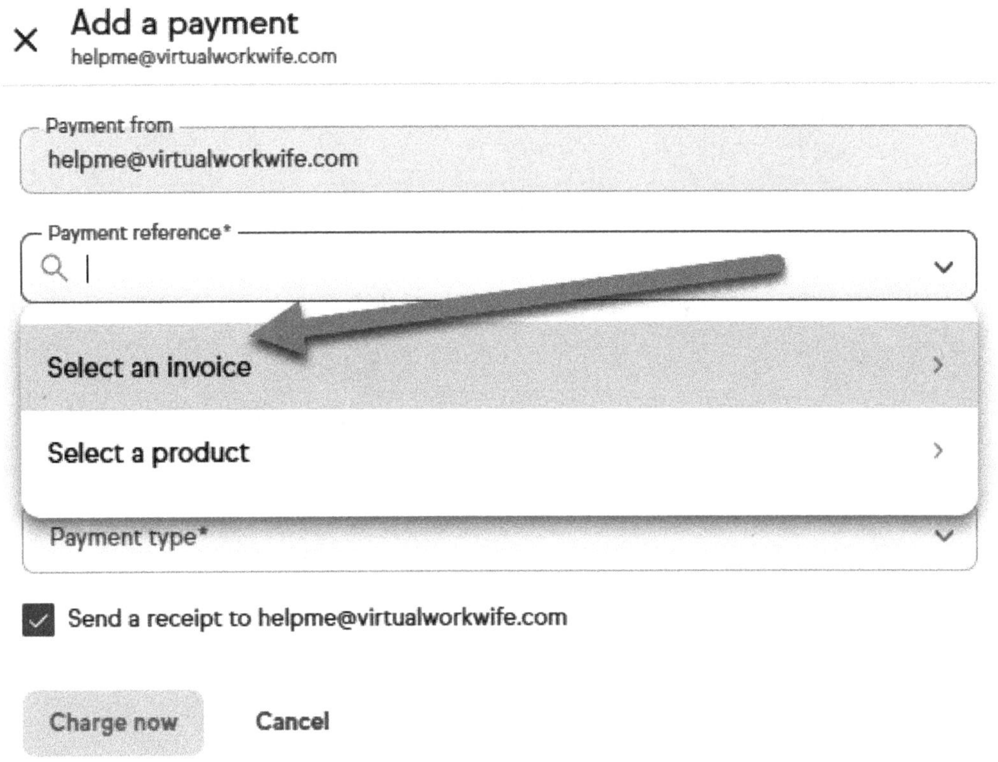

Figure 10.6 – Posting payments to an invoice

4. Use the **Payment type** dropdown to indicate what kind of payment you're recording:

    A. **Credit card**

    B. **Cash**

    C. **Check**

5. Checking the **Send a receipt to...** box will automatically send a copy to the displayed email address.

6. Click the **Charge now** button to complete your transaction.

Sending invoices and quotes

Follow these steps:

1. Using the navigation or search feature, locate the contact record you want to document a payment for.

2. You have two options for accessing quotes and invoices:

    A. Click on the **Sales** card to open the sidebar to view existing invoices and quotes.

    B. Click **More** to open the dropdown and choose **Add payment**, **Invoice**, or **Quote** to create a new one.

    For this exercise, we'll click on the **Sales** card.

3. Invoices and quotes will be flagged as **Sent**, **Paid**, or **Draft**:

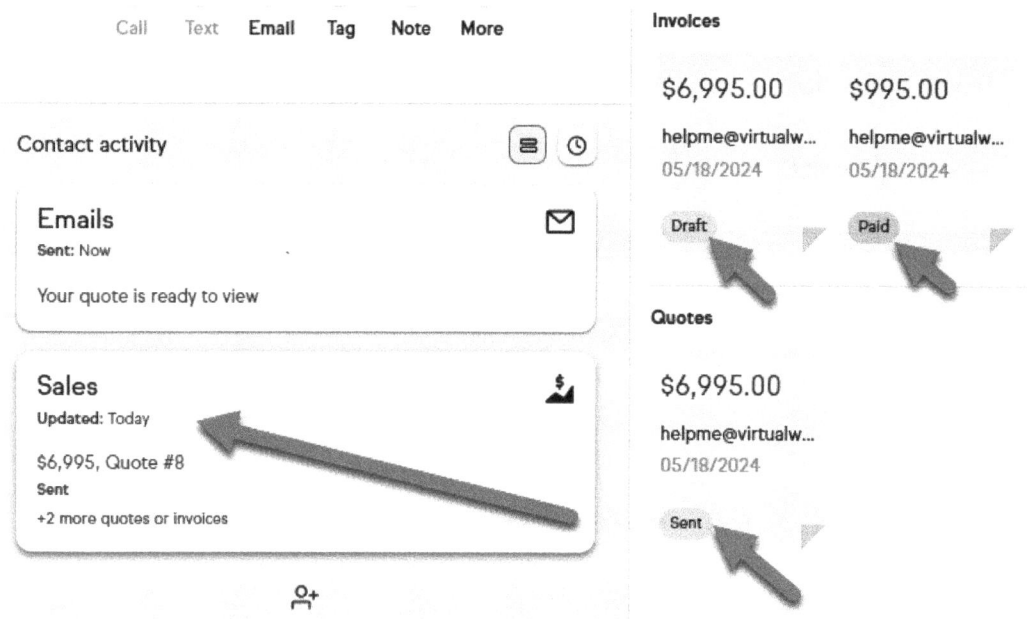

Figure 10.7 – Sending invoices and quotes

4. Click on the invoice you want to send.

5. In the top-right corner, click the **Next** button.

6. The email editor pop-up will open. Review the content of the email and click **Send**.

How it works...

When it comes to managing finances in Keap, being able to handle payments and invoicing manually is essential. Even though automated systems take care of much of the heavy lifting, having the option for manual payments allows you to swiftly record transactions, whether you're receiving cash from a neighbor or taking a client's credit card details over the phone. Plus, the invoicing feature makes billing clients a piece of cake, ensuring transparency and professionalism every step of the way. With these tools right at your fingertips, accounting will not be the dreaded chore it once was!

Manual automations

Automations are fantastic for guiding contacts along your funnel autonomously. But there are instances where a personal touch is needed. Say, for instance, you're meeting a potential client for coffee. No automation can gauge if the client shows up without a little human intervention. That's where Keap steps in, providing you with numerous ways to lend a manual hand in ushering people closer to becoming devoted buyers and fans.

How to do it...

To manually intervene and assist contacts in advancing through your funnel, start by identifying key touchpoints where human interaction may be needed. Knowing when you need to act is critical to successfully keeping your automation flowing.

In this recipe, we'll assume you've been sending automated emails, asking your contact to meet for coffee. Today, they showed up and you had a great talk.

Removing a contact from an easy automation

Here are the steps to manage contacts in an easy automation:

1. Using the navigation or search feature, locate the contact you want to move.
2. Click on the **Automations** card to open the sidebar, then click the **...** (ellipsis) icon to open the actions menu:

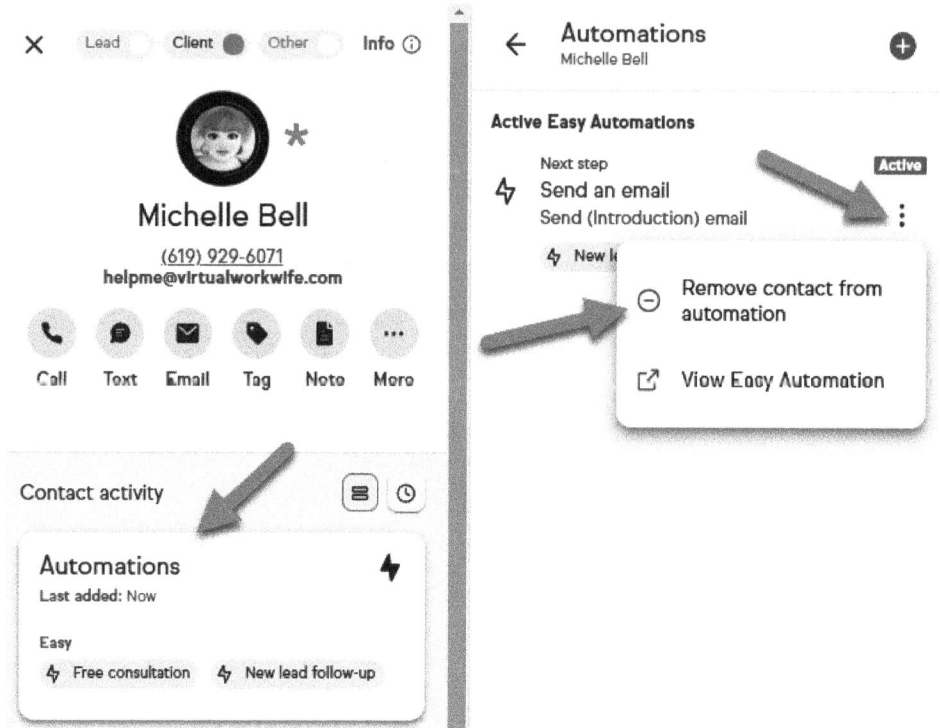

Figure 10.8 – Remove contact from automation

3. Locate the automation you want to remove the contact from and click **Remove contact from automation**.

4. Review the contact's name in the confirmation box and click **Remove** to continue.

Adding a contact to an advanced automation

Follow these steps:

1. If the contact is active in any automations, you will see an **Automations** card in their activity feed. Click to open the **Automation** sidebar.

2. Click the **More** icon to open the dropdown and choose **Add to automation**.

3. In the **Automations** sidebar, click the blue + icon in the top-right corner:

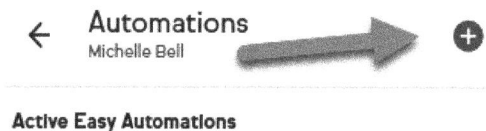

Figure 10.9 – Adding a contact to an automation

4. In the pop-up box, select the automation you want to start.

5. Select the sequence in the automation where you want to add the contact: **bronze**, **gold**, or **silver**.

6. Click **Add** to start the automation:

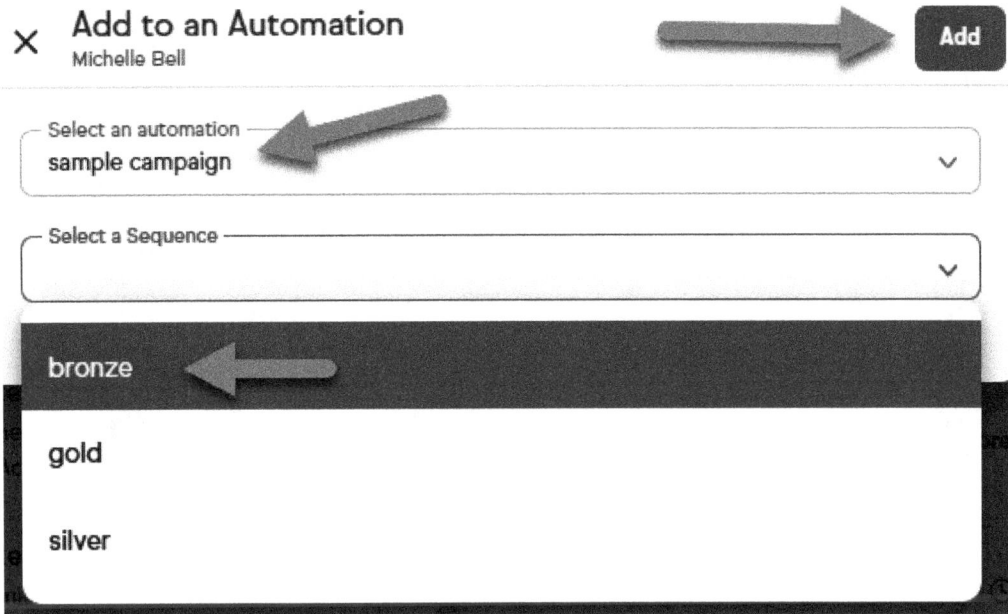

Figure 10.10 – Choosing a sequence in the automation

Simplifying the process

By utilizing tags as triggers in your automation, you can start and stop multiple automations with a single step. For our scenario, we'll use a tag called **Attended** to stop our previous automation and start a new one. Let's get started:

1. Using the navigation or search feature, locate the contact record you want to apply a tag to.

2. Click the **Tag** icon.

3. Use the search box or scroll through the features to locate the **Attended** tag.

4. Alternatively, you can click the **Add** button to create one:

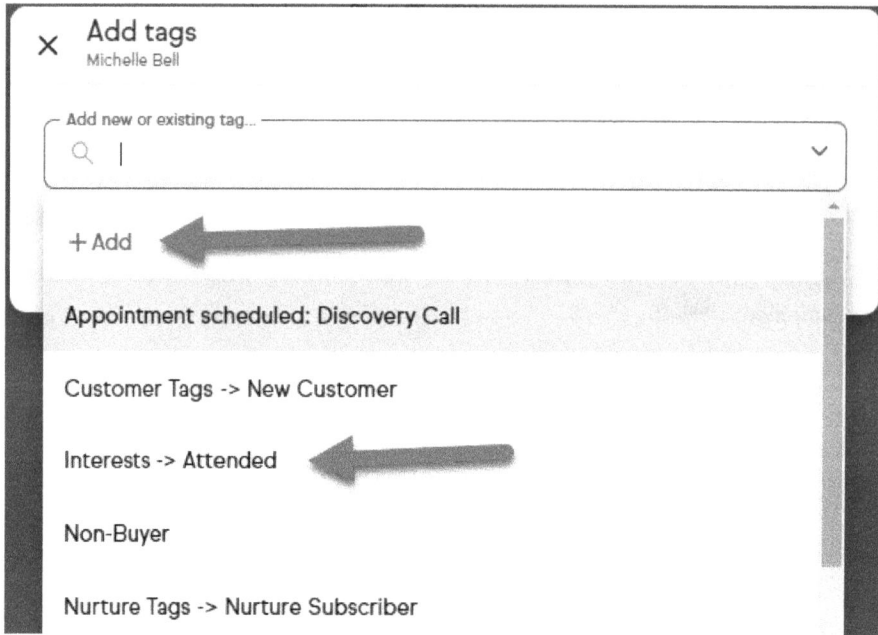

Figure 10.11 – Selecting a tag to start/stop an automation

5. Click the **Add tag to [name]** button to apply it:

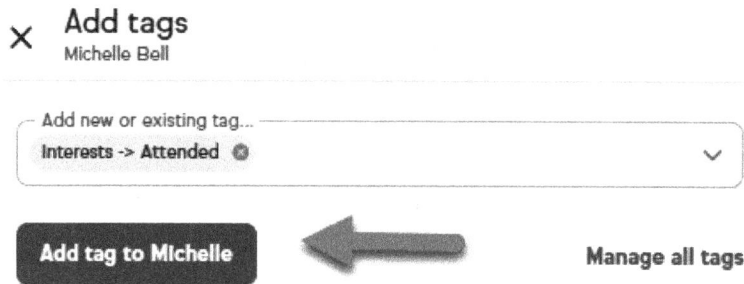

Figure 10.12 – The Add tag to [name] button

How it works...

Recognizing when you need to infuse human touch into your automated funnels is crucial to crafting personalized engagement and guiding contacts toward conversion. By pinpointing those pivotal touchpoints, setting up timely alerts, and personally connecting with contacts, when necessary, you can seamlessly marry automation with authentic human interaction. With Keap's adaptable tools and your proactive mindset, you possess the power to nurture connections, foster trust, and ultimately transform leads into devoted customers and superfans.

Creating a deal manually

We talked about deals in *Chapter 5*. So, we already know they are a big part of your overall automation game plan. But here's a little nugget of wisdom, friends. Sometimes, you've got to roll up your sleeves and manually add a contact to the deal pipeline.

Yep, that's right. Despite all the automation magic, there are moments when a personal touch is needed. So, don't hesitate to dive in and add those contacts manually when necessary. It's all part of keeping your pipeline finely tuned and ready for action.

How to do it...

To manually intervene and assist contacts in advancing through your funnel, start by identifying key touchpoints where human interaction may be needed. Knowing when you need to act is critical to successfully keeping your automation flowing.

Adding a deal to a contact

Follow these steps:

1. Using the navigation or search feature, locate the contact you want to move.

2. Click the **More** icon and select **Create deal with contact**:

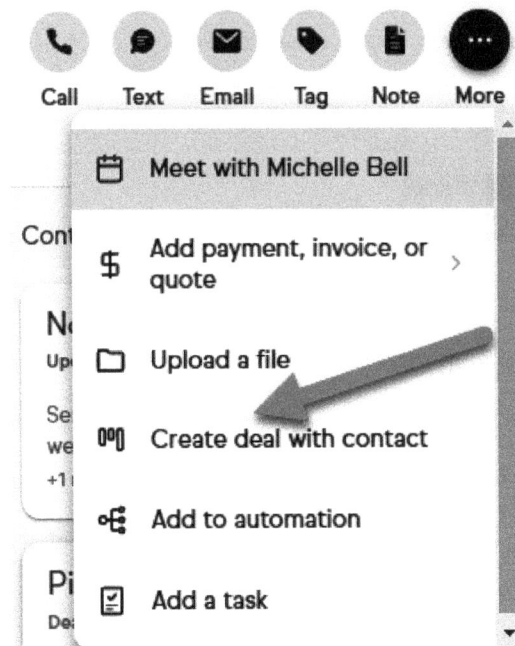

Figure 10.13 – Adding a deal to a contact

3.   In the pop-up box, select a pipeline to access the stages list.

4.   Choose a stage to start your deal at.

5.   Give your deal a name.

6.   Set the deal's value:

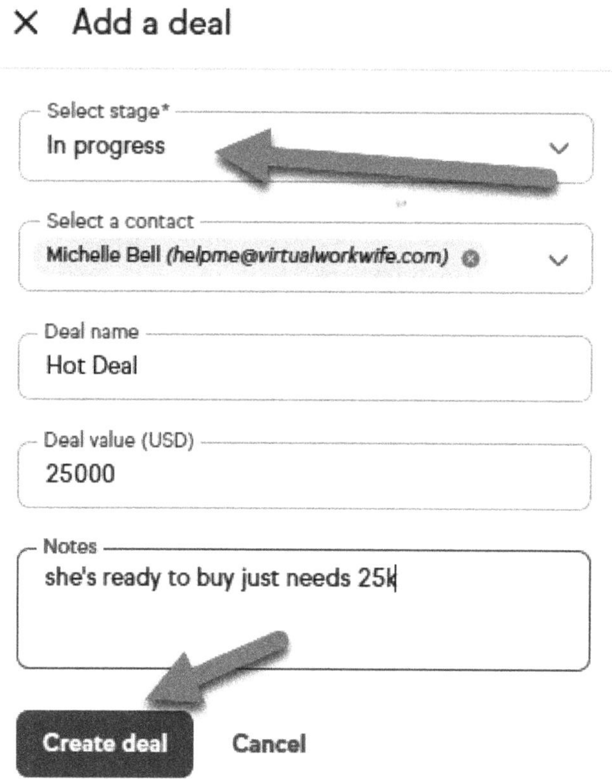

Figure 10.14 – Adding deal details

7.   Add any notes about your deal.

8.   Click **Create deal** to save it.

Manually moving a deal forward

Follow these steps:

1.   Click the **Pipeline** card to open the sidebar.

2.   Locate the deal you want to move.

3. Click the **Move to next stage** link to move the deal forward:

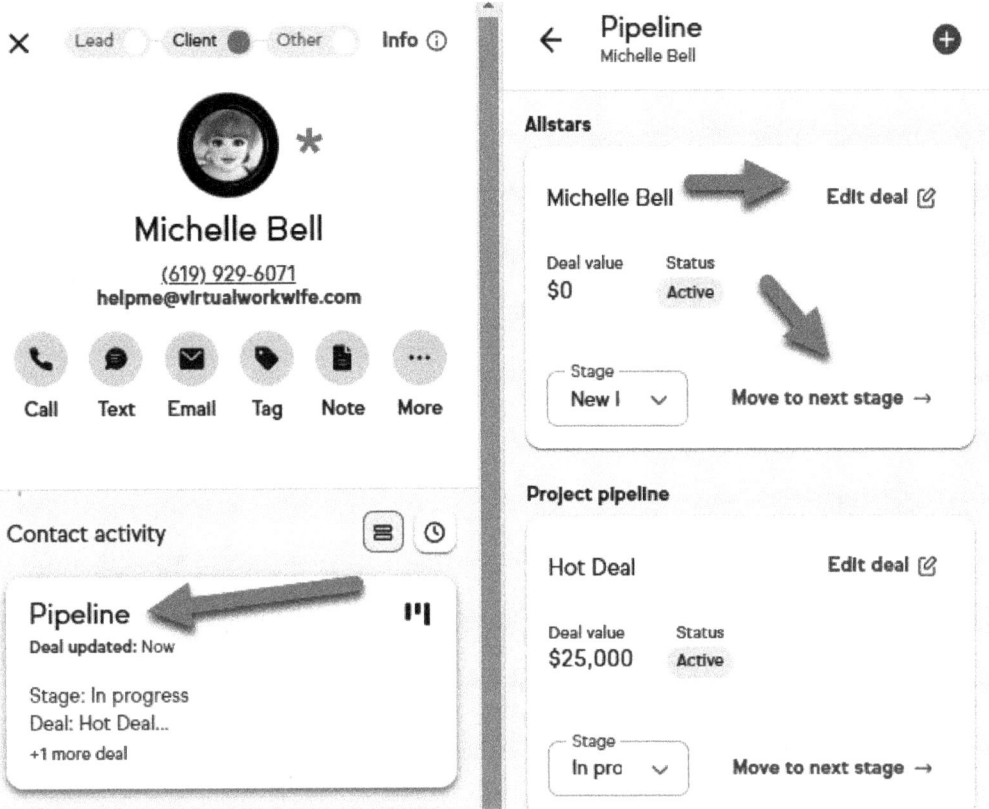

Figure 10.15 – Changing deal values

Moving a deal backward, changing the pipeline, or deleting it

Follow these steps:

1. Click the **Edit deal** link to open the deal card.

2. Click the **Stage** dropdown to select a new pipeline or stage.

3. Click the **Close** button to save your changes:

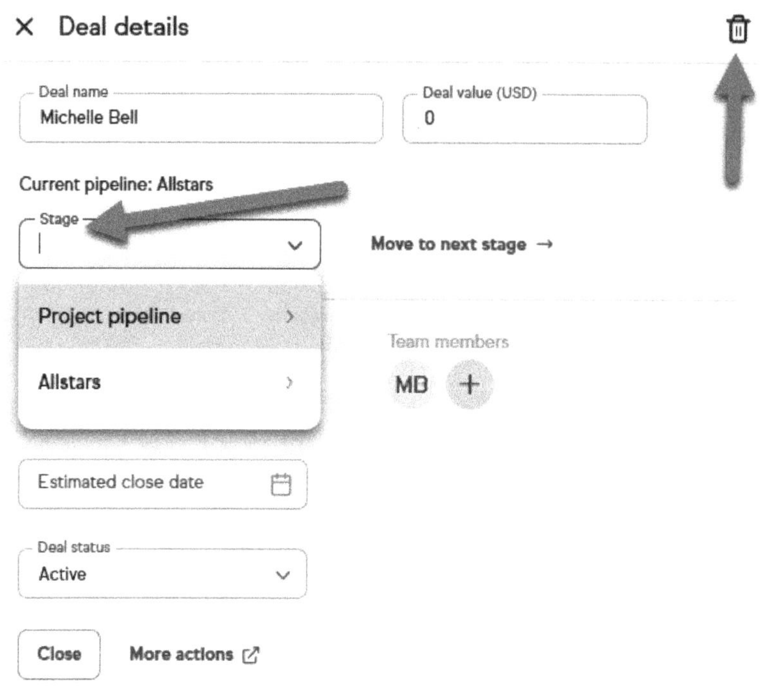

Figure 10.16 – Changing or deleting a deal

4.   If necessary, use the **trashcan** icon to delete your deal.

How it works...

When you manually edit a deal, you're stepping up and seizing the reins by updating or inserting information right into your CRM system. This guarantees precise and customized handling of every opportunity that comes your way.

But let's not stop there – this is where things get fascinating. Deals are not just about sealing the deal; they are about scrutinizing your performance from every angle. Win rates, deal sizes, conversion times – immerse yourself in that data. Find those trends, highlight your strengths, and pinpoint those areas where you can level up. It's all about refining your strategies to make a lasting impact.

Using notes

Let's talk about why jotting down notes in your CRM is an absolute game-changer. First off, they are like your trusty memory bank, holding all the nitty-gritty details of your interactions and connections. This means you're not just shooting in the dark; you're armed with insights into what makes each person tick, helping you tailor your approach to their needs and preferences like a pro.

But wait, there's more! Notes are not just for your eyes only; they are the secret sauce for team synergy. When you're all on the same page with updates, insights, and to-dos neatly documented, you can bet your bottom dollar that your customer service will be smoother than silk, no matter who's handling it.

How to do it...

To manually intervene and assist contacts in advancing through your funnel, start by identifying key touchpoints where human interaction may be needed. Knowing when you need to act is critical to successfully keeping your automations flowing.

Adding a note to a contact

Follow these steps:

1. Using the navigation or search feature, locate the contact you want to update.

2. Click on the **Notes** icon to open the pop-up box.

3. Add your note text in the box and click the **Save** button:

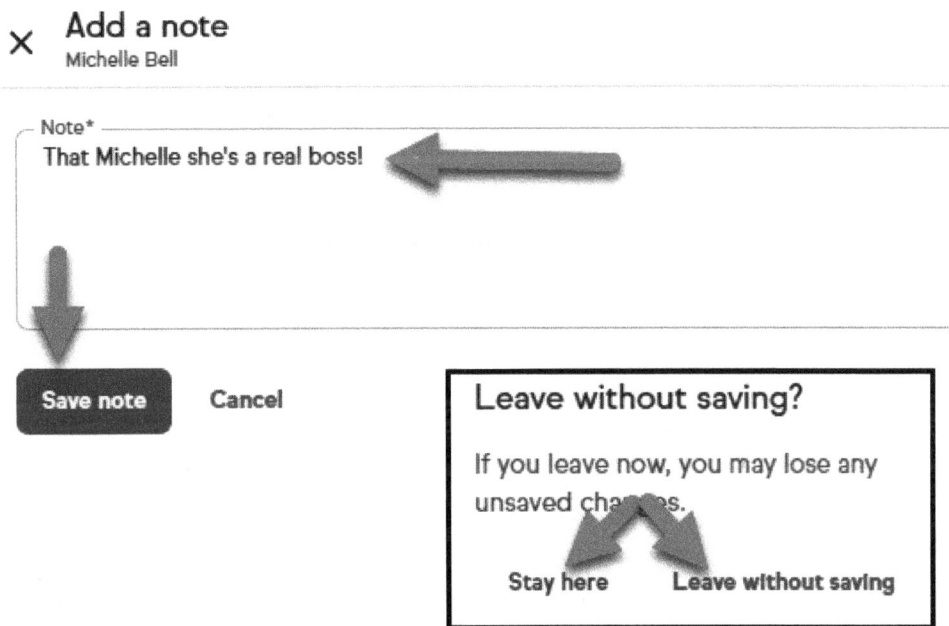

Figure 10.17 – Adding notes to a record

> **Note**
> If you click outside the note before saving it, a warning box will ask you if you want to stay or leave without saving.

Editing or deleting a note

It's easy to edit or delete your notes after the fact. Let's get started:

1. Using the navigation or search feature, locate the contact you want to update.

2. Click on the **Notes** icon to open the sidebar.

3. Click on the note you want to edit to open the pop-up box.

4. Make your changes and select **Save changes** or **Delete**:

    A. If you're deleting, review the confirmation box and select **Delete this note** to continue:

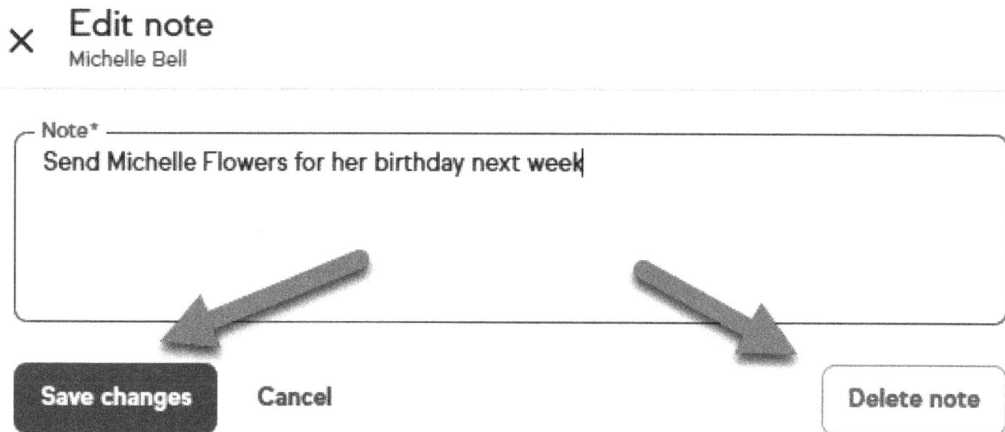

✕ **Edit note**
Michelle Bell

Note*

Send Michelle Flowers for her birthday next week

[Save changes]   Cancel   [Delete note]

Figure 10.18 – Choosing a sequence in the automation

How it works...

Taking full advantage of notes in your CRM isn't just a good idea – it's a total game-changer. It amps up your efficiency, amps up your teamwork, and, most importantly, amps up your customers' satisfaction. So, what are you waiting for? Get those notes flowing and watch your relationships thrive!

Using tasks

Notes are super handy for capturing insights, holding onto essential details, and fostering collaboration. And tasks? Well, they are like the busy worker bees, buzzing around our CRM with purpose. Each one represents a specific action or activity that needs to be done within a certain timeframe.

Why are they so crucial? Because they are the key to staying organized, prioritizing what needs to be done, and making sure everyone is accountable for their part in the workflow. They are the engine that keeps things moving forward, whether it's sealing deals, completing projects, or engaging with customers.

How to do it...

Tasks are the backbone of your productivity. So, when the need arises to create a manual task related to a contact, here's what you will need to do.

Adding a task to a contact

Follow these steps:

1. Using the navigation or search feature, locate the contact you want to update.
2. Click on the **More** icon to open the drop-down menu.
3. Select **Add a task**:

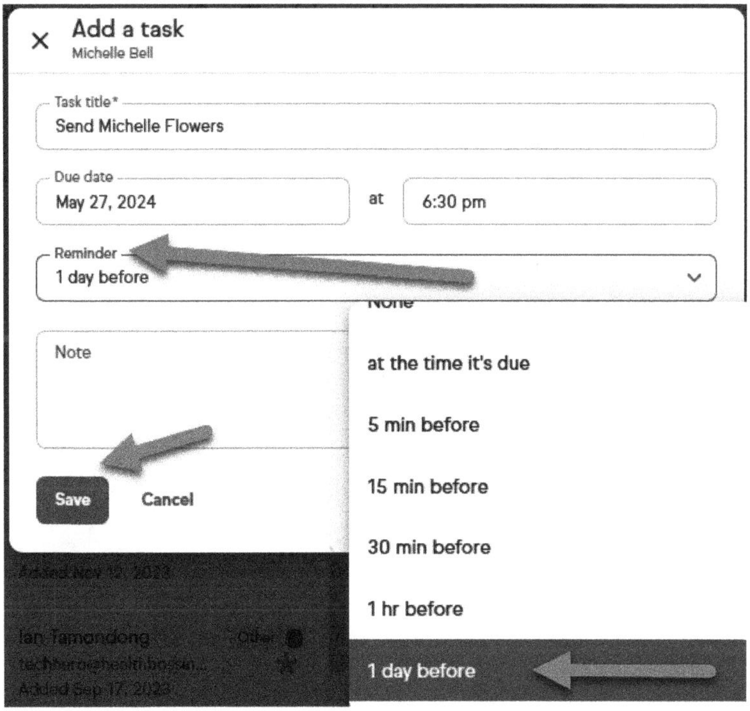

Figure 10.19 – Creating a manual task

4. Give your task a title.
5. Set a due date and time.
6. Click the **Reminder** dropdown to select an option.
7. Add any additional notes.
8. Click **Save** to close this task.

Editing or deleting a task

Follow these steps:

1. Click on the **Task** card to open the sidebar, then click on the task you want to edit.

2. Make any changes needed and click **Save**.

3. Alternatively, you can click the **Delete** button to remove it:

    A. If you're deleting the task, a confirmation box will appear.

    B. Click **Delete** to continue:

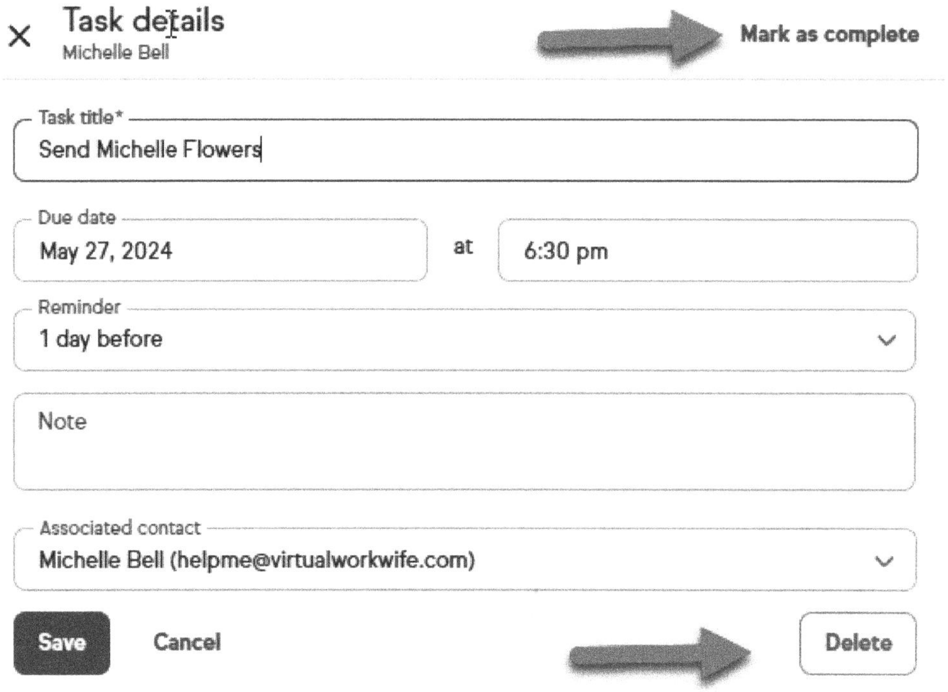

Figure 10.20 – Editing or deleting a task

Marking a task as complete

Marking tasks as complete ensures you have the most accurate information on your contact records. Follow these steps:

1. Click on the **Task** card to open the sidebar, then click the task you want to edit.

2. Click the **Mark as complete** link to update the task. Click on the note you want to edit to open the pop-up box.

3. Make your changes and select **Save changes** or **Delete**:

A. If you're deleting, review the confirmation box and select **Delete this note** to continue:

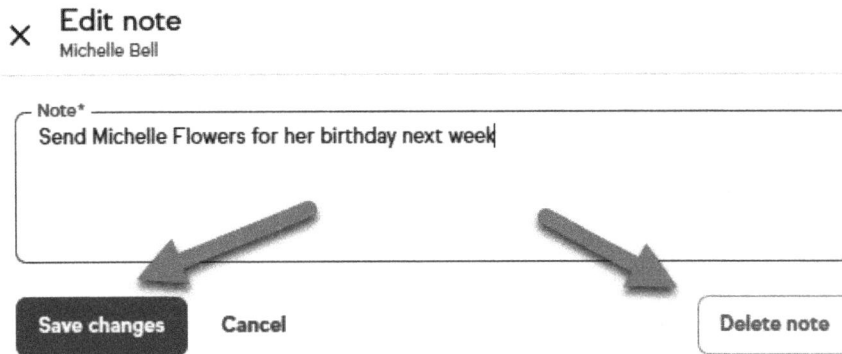

Figure 10.21 – Choosing a sequence in the automation

How it works...

Imagine tasks as your trusty guide through the maze of deals, projects, and customer interactions. They are not just boxes to check off; they are the backbone of your productivity. Tasks make sure nothing falls through the cracks and that every action is taken with purpose. So, embrace your notes and tasks – they are the dynamic duo that keeps your CRM humming along smoothly!

Requesting Google Reviews

Here's the scoop: the more reviews you rack up, the more leads come knocking at your door. And guess what? Google Reviews just made it a breeze to get those glowing testimonials. It allows you to boost your online visibility, keep tabs on who's singing your praises, and shoot off customized messages to reel in even more reviews. It's a win-win for your business growth and reputation.

How to do it...

When it comes to obtaining reviews, or anything in life, the answer is always no if you don't ask! Follow these steps to make sure you're getting all the yesses.

Requesting a review

Follow these steps:

1. Navigate to your home page (dashboard) and find the **Google Reviews** widget.

2. Click the **Request review** button to open the email editor.

3. Use the search box to add your recipient to the email.

4. You can edit the default email copy if needed.

> **Note**
>
> Do not edit or change the request link, as shown in *Figure 10.21*. Doing so may prevent recipients from being able to leave you a review!

5. Use the **Signature** toggle to include your contact details if desired.

6. Click **Send**:

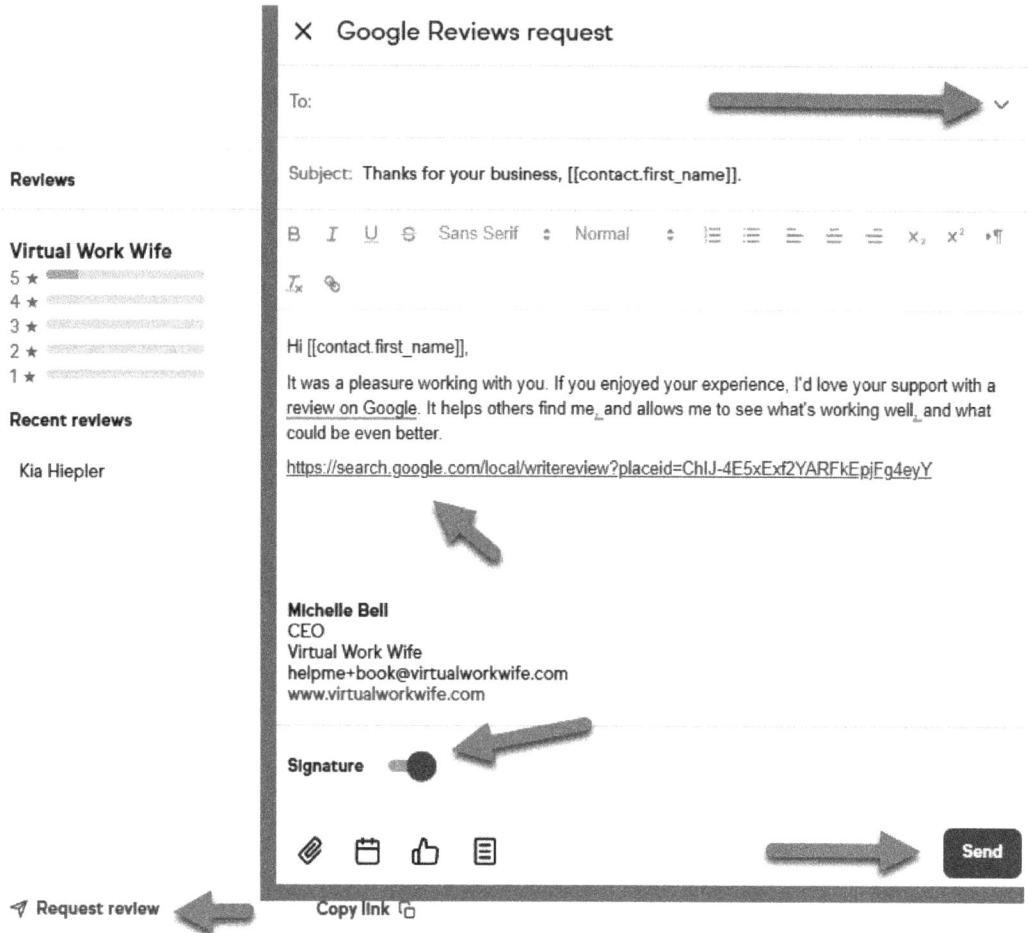

Figure 10.22 – Requesting a Google review

> **Pro tip**
>
> Want to up your chances of snagging those coveted Google reviews on the regular? Here's a handy trick: grab the link from your widget and pop it into an email template. Then, set up an automation to do the heavy lifting for you. Whenever a client wraps up a transaction, voilà! Your request for a review zips off automatically. It's like having a personal review-generating machine!

How it works...

Harnessing the full potential of your CRM to stay on top of those Google reviews is critical. In today's digital age, your online reputation holds immense sway over the success of your business. Google reviews are often one of the first things potential customers see when they search for your business. Positive reviews can act as powerful endorsements, influencing people's purchasing decisions and ultimately driving more business your way.

By strategically encouraging Google reviews using savvy methods such as incorporating review links into widgets, crafting bespoke email templates, implementing automated review requests, and tactfully following up, you're not just amassing feedback; you're cultivating trust and credibility among your audience.

Keap is your linchpin. It streamlines these processes, allowing you to efficiently track and manage customer interactions. This means you can easily pinpoint who you have contacted for reviews and who still warrants a gentle nudge.

Managing your daily routine

In Keap, your dashboard acts as the ultimate command center, offering a holistic view of your business's vital stats and operations. From tracking sales performance to monitoring customer interactions and marketing campaigns, it's your go-to for real-time insights and informed decision-making. With all your crucial data neatly organized in one intuitive interface, Keap's dashboard equips you to streamline operations, fine-tune strategies, and propel your business toward growth and success.

Customizing your dashboard in Keap is a breeze, allowing you to tailor it to suit your unique business needs and preferences. Whether you want to rearrange widgets, add new metrics, or adjust the layout, Keap offers intuitive tools that make customization a cinch.

With just a few clicks, you can personalize your dashboard to focus on the metrics that matter most to you, ensuring that you have the insights you need at your fingertips. The following dashboard widgets are available in Keap:

- **Contacts**: Tracks the current count of all your contacts within Keap.
- **New Leads**: Tracks contacts that have been tagged as **Lead**. Use the drop-down filter to choose a timeframe for tracking. Note that the tag application date and the date a contact is added to Keap may not be the same.

- **New Clients**: Tracks the influx of new clients by tallying contacts tagged as **Client** within your specified timeframe. Remember, we are tracking when they were tagged, not when they were first added to Keap.

- **Repeat Clients**: Keep an eye on customer loyalty by observing how many clients returned for another purchase within a specified timeframe.

- **Sales**: See your recent sales performance and compare it with the previous period to gauge growth or changes.

- **Quotes**: Monitor the number of quotes you have sent out in the last month and how many were accepted.

- **Invoices**: Keep track of your revenue by monitoring the total amount invoiced and the revenue received from paid invoices over the last 30 days.

- **Broadcasts**: Stay informed about the open and click rates of your latest email broadcast to assess engagement levels.

- **Recent Activity**: Displays a list of the last ten interactions you've had within Keap. These are clickable, making it easy for you to quickly revisit recent items.

- **Tasks**: Displays a list of tasks assigned to you, sorted by due date.

- **Appointments**: Displays contact information along with appointment details for upcoming appointments scheduled within the next week.

- **Email Health**: Provides valuable and actionable insights to assist you in optimizing email deliverability to your contacts' inboxes.

- **Reviews**: Displays a Google review score and information on recent reviews.

How to do it...

This recipe will walk you through customizing your dashboard so that you can make it relevant to your daily needs.

Hiding or displaying dashboard widgets

Follow these steps:

1. Navigate to your home page.
2. Click the ... (ellipses) icon on any widget.
3. From the drop-down menu, select **Manage widgets**:

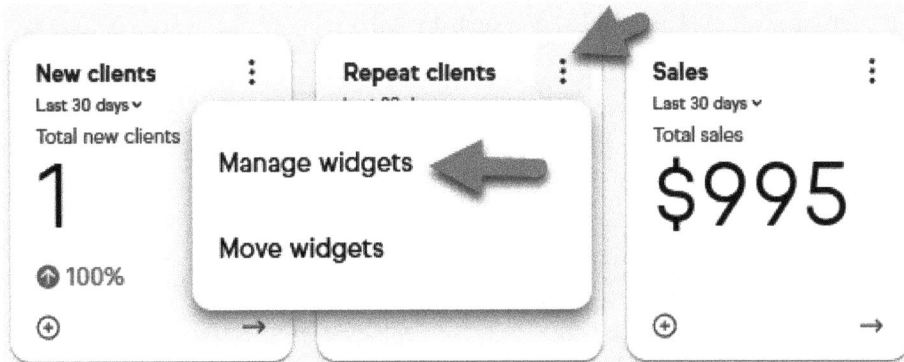

Figure 10.23 – Managing dashboard widgets

4. In the pop-up box, use the toggle buttons to turn your desired widgets on or off:

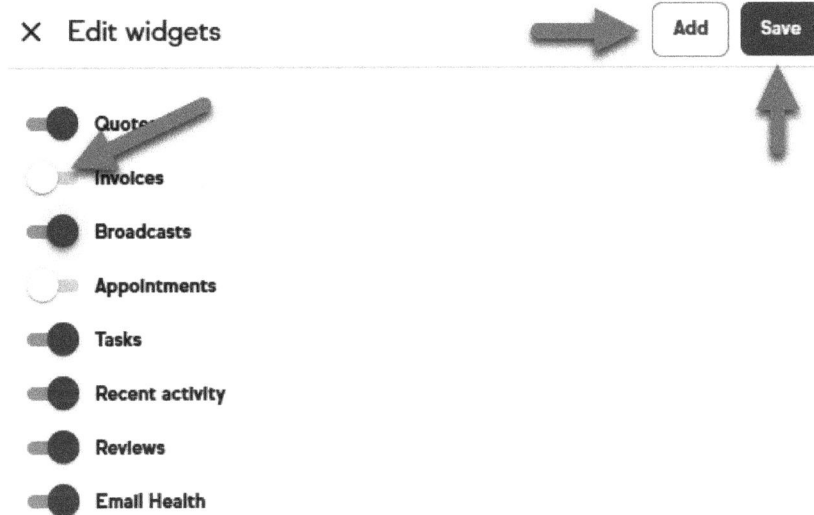

Figure 10.24 – Turning widgets on/off

5. Click **Save**.

Adding group widgets

You can add groups of contacts to a widget by using filters such as tag, regions, creation date, and more. Let's get started:

1. Click the **...** (ellipses) icon on any widget.
2. Click the **Add** button.

3. In the **Groups** box, click the **Add to dashboard** + link:

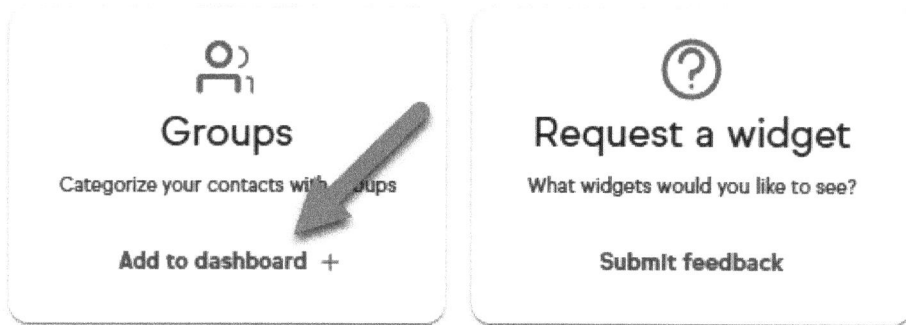

Figure 10.25 – Adding groups

4. Add a name for your group:

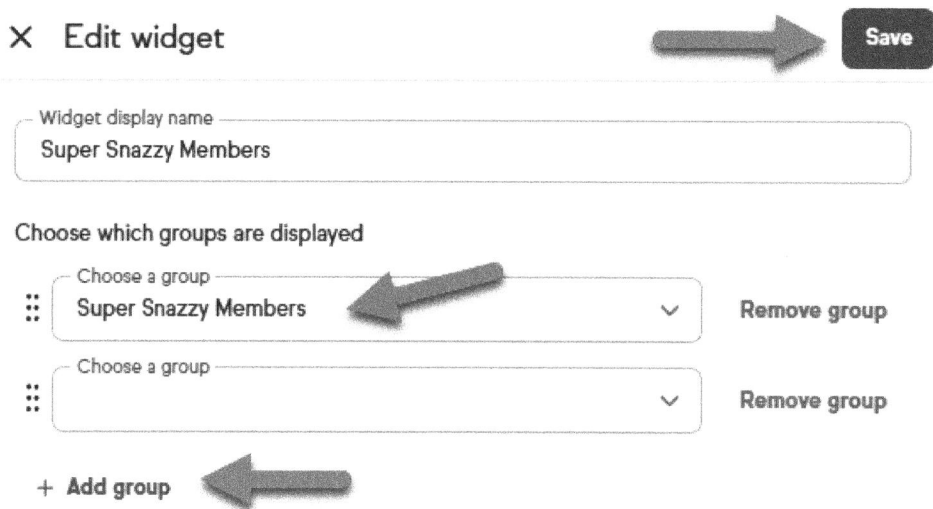

Figure 10.26 – Adding group filters

5. Choose an existing group from the dropdown or click the + **Create a new group** link to create a new one.

6. To add another group, click the + **Add group** link and repeat *Step 4*.

7. Repeat *Steps 4* and *5* until you are satisfied with your additions.

8. Click **Save**.

Arranging widgets on your dashboard

You can easily move your widgets to arrange them in a way that works best for you. Follow these steps:

1. Click the ... (ellipses) icon on any widget.

2. Click the **Move** button.

3. Using your mouse click and hold on any widget to drag it to another area of your dashboard.

4. Click the **Done** button in the top-right corner to save your new dashboard arrangement.

How it works...

Your dashboard serves as the heart of your business – it's the ultimate tool for effortlessly managing your daily activities!

Packed with everything you need, it offers intuitive data visualization, real-time updates, and customizable layouts tailored to your preferences. No more searching for scattered information or juggling multiple tools – this powerhouse consolidates it all into one convenient location.

And why does it matter? Because you are all about driving progress, my friend. With this dashboard at your disposal, you are not just keeping track – you're capitalizing on opportunities at every turn. Identify trends, monitor progress, and take action like never before. Say goodbye to wasted time and hello to supercharged productivity.

Utilizing the My day page

The **My day** page simplifies your workflow by consolidating tasks and appointments into one convenient page. With **My day**, you can efficiently manage your daily activities within a focused area of Keap, eliminating the need to jump back and forth between various pages within your CRM.

How to do it...

From **My day**, you can manage all of your appointments and tasks in one consolidated location. In this recipe, we will cover how to navigate both views.

Managing appointments

Follow these steps:

1. Navigate to the **My day** page by clicking the icon on the navigation bar.

2. Choose **Appointments** from the submenu.

3. You can now do the following:

    A. View all your booking links.

    B. Add new booking links by clicking the + icon.

    C. Scroll through your calendar using the <> arrows.

    D. Easily send booking links to contacts from the dropdown.

    E. Manually add appointments to your calendar using the **Book now** button.

    F. Filter your calendar.

    G. Click on the entries to open appointments and view details:

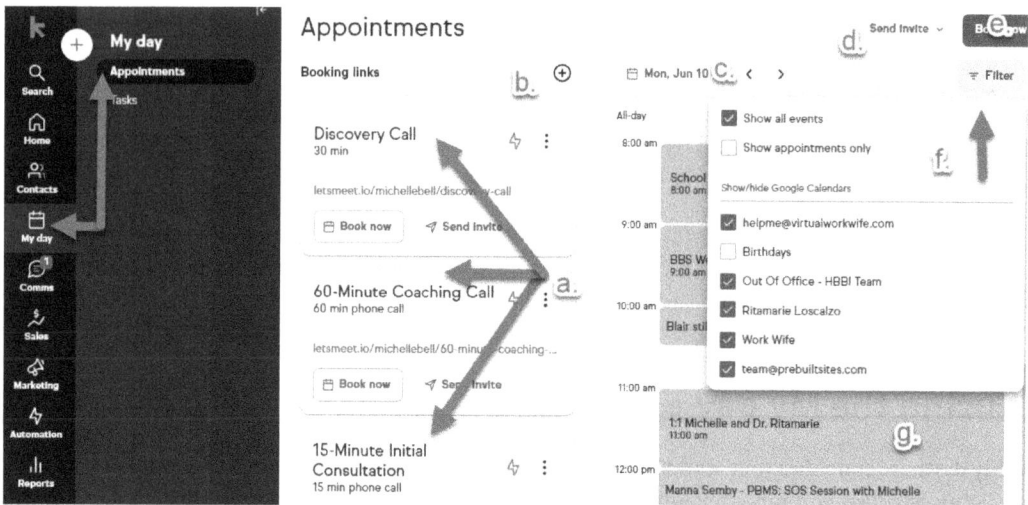

Figure 10.27 – Navigating the My day Appointments view

Managing tasks

Follow these steps:

1. Navigate to the **My day** page by clicking the icon on the navigation bar.

2. Choose **Tasks** from the submenu.

3. You can now do the following:

    A. View past due, current, and future tasks.

    B. View completed tasks (hidden by default).

    C. Manually add tasks.

D. Click on the tasks to open them and update their details:

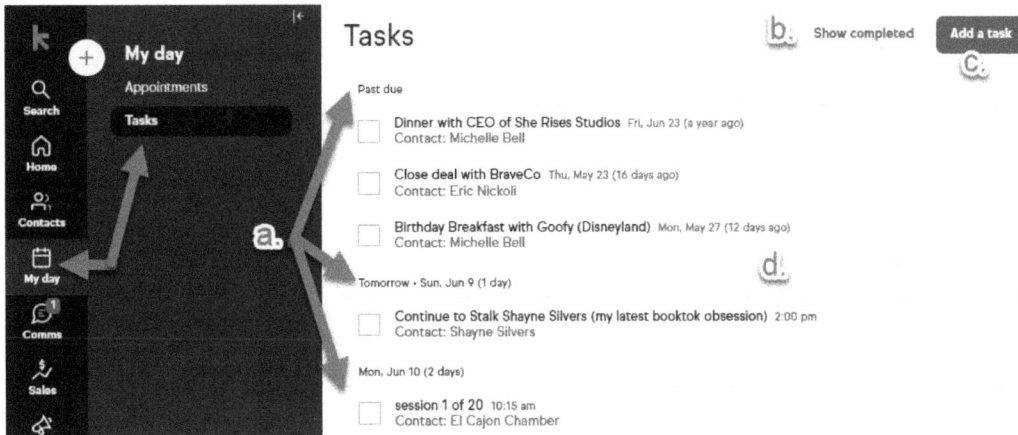

Figure 10.28 – Navigating the My day appointments view.

How it works...

Having all your appointments and tasks in one view within Keap can be a true paradigm shift for organization and stress management.

It streamlines your day, saving you valuable time that would otherwise be spent toggling between different calendars and lists. With everything consolidated, you can easily spot scheduling conflicts, plan more effectively, and avoid the anxiety of overbooking or overloading team members.

Plus, it's not just about you; having a unified view fosters better communication and coordination within your team, ensuring everyone is on the same page.

Overall, this approach enhances efficiency, promotes better time management, and provides the peace of mind needed for a stress-free day.

11

Five Essential
Automation Funnels

In *Chapters 1* through *10*, we covered the *how* of Keap's inner workings. Now, it's time to get down to business. Whether you're new to the online business scene or a seasoned entrepreneur, linking your CRM to your website is a must.

In today's digital world, every interaction matters. Efficiently capturing and managing leads can make or break your success. Your website serves as your digital storefront, attracting potential customers with valuable information. But without CRM integration, you're missing out on opportunities to nurture leads and turn them into loyal customers. In short, you're leaving money on the table.

Consider this: A visitor lands on your site, intrigued. They explore your content, maybe fill out a form or subscribe. Without integration, this data might get lost or buried. But with Keap's automation, you can capture leads in real time, organize them, and personalize follow-ups to guide them through the sales journey.

Imagine: A prospect fills out a form, expressing interest. With Keap integrated into your site, their info is instantly in your CRM, triggering tailored follow-ups. They get a thank-you email with a link to schedule a call, and they're added to your newsletter for ongoing engagement.

The sooner you act, the sooner you streamline your processes and ditch other platforms and redundant software services.

To speed up your implementation, in this chapter, we've mapped out five essential funnels:

- Newsletter opt-in
- **Contact Us**
- Discovery call
- Lead magnet
- Pay fail

Technical requirements

For this chapter, the following skills may be helpful:

- Basic website editing skills—the goal of these automations is to connect your website to your CRM in order to grow your list and serve your audience in a more efficient way

Newsletter opt-in

A **newsletter opt-in** is a feature on a website that allows visitors to subscribe to a newsletter. Typically, it involves a form where visitors can enter their email address and sometimes additional information, such as their name or interests. By opting in, visitors are giving permission to receive regular updates, news, promotions, or other content from the website or business via email. This allows businesses to build and maintain a direct line of communication with interested individuals and potential customers.

For a basic newsletter opt-in, you're going to need three things:

1. A form to collect data
2. A welcome/confirmation email
3. A tag to segment the group

How to do it...

1. Follow the instructions in *Chapter 6* to build a public form that asks for the following:

   A. First name

   B. Email

2. Style your form using your branding and design preferences.

3. Automate your follow-up by following the instructions in *Chapter 7*. Create an automation using *when* and *then*:

   When the newsletter form is submitted

4. *Then* apply a tag.

5. *Then* send an email confirming you received their request and setting expectations for what happens next.

6. Decide on what kind of thank-you page you want to use. As discussed in *Chapter 6*, you can choose from either of the following:

   A. A Keap-hosted form that you design

   B. Redirect to a URL

7. Test your form!

    A. Are there any typos?

    B. Did you get tagged?

    C. Did you receive the email?

    D. Do links in the email work?

    E. Did you land on the right thank-you page?

8. Implement the form in your workflow. Remember that you have the option to do the following:

    A. Use the embed code to place the form on your website

    B. Use the hosted link to connect it to a button

    C. Share on your social channels

9. Following the instructions in *Chapter 10*, add a newsletter tag to a widget on your dashboard.

How it works...

Adding a newsletter opt-in to your website promptly is crucial because it establishes a direct line of communication with visitors, allowing you to cultivate relationships, share valuable content, and promote your business or brand effectively. It enables you to build a subscriber base, nurture leads, and drive engagement, ultimately contributing to the growth and success of your online presence.

Contact Us

A **Contact Us** form is often the gateway for visitors to connect with a company or organization, whether they have inquiries, feedback, or need assistance. Surprisingly, it's frequently disregarded in terms of workflow automation, yet it's crucial for ensuring all leads are captured in your CRM!

For a basic **Contact Us** form, you're going to need four things:

- A form to collect data
- A confirmation email
- A tag to segment the group
- A task to follow up with the contact

How to do it...

1. Following the instructions in *Chapter 6*, build a public form that asks for the following:

    A. First name

    B. Email

    C. Additional comments—so that they can share their question(s) with you

2. Style your form using your branding and design preferences.

3. Automate your follow-up. Following the instructions in *Chapter 7*, create an automation in the following sequence:

    A. *When* the **Contact Us** form is submitted

    B. *Then* apply a tag.

    C. *Then* send an email, confirming you received their request and setting expectations for what happens next. (This is a great time to also ask them to follow you on social media!)

    D. *Then* create a task to follow up.

4. Decide on what kind of thank-you page you want to use. As discussed in *Chapter 6*, you can choose from either of the following:

    A. A Keap-hosted form that you design

    B. Redirect to a URL

5. Test your form!

    A. Are there any typos?

    B. Did you get tagged?

    C. Did you receive the email?

    D. Do links in the email work?

    E. Did you land on the right thank-you page?

6. Implement the form in your workflow:

    A. Use the embed code to place the form on your website.

    B. Use the hosted link to connect it to a button.

    C. Share on your social channels.

How it works...

Having a **Contact Us** form on your website that syncs seamlessly with your CRM system is an absolute must for any business serious about cultivating a genuine customer relationship experience. It's like having a direct line to your customers' needs and inquiries, all neatly organized in one place. By connecting the form to your CRM, you're not just capturing contact details; you're capturing valuable insights into customer behavior and preferences, which can inform your marketing and sales strategies.

Plus, with this integration, you're not just collecting data—you're automating processes too. Assigning leads, tracking interactions, and sending personalized follow-ups become a breeze rather than a burden. It's all about efficiency and delivering top-notch customer service. So, if you're looking to streamline communication, understand your customers better, and boost your business's growth, adding a CRM-connected contact form is the way to go.

Discovery call

When it comes to sales, you can't go wrong with a good discovery call form. By asking targeted questions upfront, you position yourself for a smoother call and greater success. Not only does it make your call more efficient, but it also makes the other person feel truly understood. With time to review their answers beforehand and employing attentive listening, you can tailor an irresistible offer perfectly suited to your ideal prospects. Plus, by utilizing this form, you can weed out the tire kickers, saving you time and hassle.

For a basic discovery call, you're going to need seven things:

1. A form to collect data
2. Four to six targeted questions to uncover their needs, challenges, goals, budget, and timeline
3. A confirmation email
4. A link to your appointment scheduler
5. A tag to segment the group
6. A task to follow up with the contact
7. Three to five follow-up emails if they don't purchase during your call

How to do it...

1. Following the instructions in *Chapter 6*, build a public form that asks for the following:

    A. First name

    B. Last name

    C. Phone number

    D.   Email

    E.   Your four to six targeted qualifying questions

2.   Style your form using your branding and design preferences.

3.   Automate your follow-up. Following the instructions in *Chapter 7*, create an automation with these steps:

    A.   *When* the discovery form is submitted

    B.   *Then* apply a tag.

    C.   *Then* send an email.

    D.   *Then* create a task to follow up.

    E.   *Then* create a deal.

4.   Add another automated follow-up. Again, following the instructions in *Chapter 7*, create an automation with the following steps:

    A.   *When* a deal moves into follow-up

    B.   *Then* send an email. Add your four to six emails.

    C.   *Stop* when a purchase is made.

5.   Set your thank-you page to redirect to your scheduling link.

6.   Test your form!

    A.   Are there any typos?

    B.   Did you get tagged?

    C.   Did you receive the email?

    D.   Do links in the email work?

    E.   Did you land on the right thank-you page?

    F.   Were you able to schedule a call?

    G.   Did you receive a call confirmation email?

7.   Implement the form in your workflow:

    A.   Use the embed code to place the form on your website.

    B.   Use the hosted link to connect it to a button.

    C.   Share on your social channels.

For more information about what emails to use and a workflow map outlining the ideal steps for a discovery call, check out the *bonus chapters* on GitHub: `https://github.com/PacktPublishing/Keap-Cookbook`

How it works...

Having a discovery call form with qualifying questions is pivotal when it comes to propelling your business forward swiftly. Picture it as a strategic filter that ensures every call you make counts, connecting you with prospects genuinely interested in what you offer. By asking the right questions upfront, you're not just saving time; you're making your calls more effective. You get a clear picture of your prospect's needs, challenges, and where they stand in their decision-making process, allowing you to tailor your pitch perfectly.

But it's not just about saving time; it's about streamlining your entire sales process. With this form in place, you're able to focus your energy on leads with the highest potential for conversion, making your efforts more targeted and productive. Plus, the insights you gather from these questions aren't just for the call—they're invaluable for refining your marketing strategies and honing in on your ideal audience. So, by integrating a discovery call form with qualifying questions, you're not just moving your business forward; you're doing it with precision, purpose, and speed.

Lead magnet

Using a lead magnet is like planting seeds for your email list. It's a powerful way to attract and engage potential customers by offering something of value in exchange for their contact information. It not only grows your list but also builds trust and credibility with your audience.

Once they've expressed interest in learning more about you by requesting your lead magnet, you have the chance to send emails (often termed *indoctrination*, though I prefer *my story* as it feels less icky) to warm them up and draw them deeper into your world, bringing them one step closer to becoming buyers.

For a basic lead magnet, you're going to need five things:

1. A landing page to collect data

2. A free giveaway such as a guide, video, worksheet, or listicle

3. A welcome/delivery email

4. A tag to segment the group

5. Three to five emails to tell your story

How to do it...

1. Following the instructions in *Chapter 6*, build a public form that asks for the following:

    A. First name

    B. Email

2. Style your form using your branding and design preferences.

3. Automate your follow-up. Following the instructions in *Chapter 7*, create an automation in the following sequence:

    A. *When* the lead magnet form is submitted

    B. *Then* apply a tag.

    C. *Then* send an email delivering your free giveaway. If you will be attaching a file to your email, follow the instructions in *Chapter 4*.

    D. *Then* send an email. Add your three to five *my story* emails.

4. Decide on a thank-you page:

    A. Keap-hosted form

    B. Redirect to a URL

5. Test your form!

    A. Are there any typos?

    B. Did you get tagged?

    C. Did you receive the delivery email?

    D. Were you able to download/open the free item?

    E. Do links in the email work?

    F. Did you land on the right thank-you page?

6. Implement the form:

    A. Use the embed code to place the form on your website.

    B. Use the hosted link to connect it to a button.

    C. Share on your social channels.

7. Following the instructions in *Chapter 10*, add the lead magnet tag to a widget on your dashboard.

For more information about suggested email themes to use and a workflow map outlining the ideal steps for a lead magnet, check out the *bonus chapters* on GitHub: `https://github.com/PacktPublishing/Keap-Cookbook`

This strategy isn't just for those new to marketing—it's a fundamental step for anyone looking to engage potential customers effectively. It's like creating a welcoming journey for your audience, enticing them with valuable content, and gradually introducing them to your brand. By implementing a lead magnet followed by a *my story* sequence, you're not only capturing their interest but also nurturing a relationship that can lead to conversions.

Pay fail

Contacting individuals promptly when their payment fails is crucial for maintaining a positive customer experience. By reaching out swiftly, you can address any issues, offer support, and help them quickly resolve the payment problem. This proactive approach demonstrates reliability and care, potentially salvaging the transaction and preserving the customer relationship while making sure you still get paid!

For a basic pay fail, you're going to need four things:

1. A tag to trigger your automation
2. A custom pipeline
3. Three to five emails
4. A task to follow up

How to do it...

1. Following the instructions in *Chapter 9*, run a failed invoice report:

   A. Check the **Select all** box.

   B. Click the **Actions** button, then select **Apply/Remove tag**.

   C. Under the drop-down box, click the **Create a tag** link and create a tag named `PayFail`.

> **Note**
> You only have to create the tag one time. In the future, you will select the tag from the drop-down box.

   D. Click **Save**.

   E. Click **Process Action**.

2. Following the instructions in *Chapter 5*, create a custom pipeline:

    A. Add a **Pay Fail** stage.

    B. Add a **Pay Success** stage.

    C. Save your pipeline.

3. Following the instructions in *Chapter 7*, create an automation in the following sequence:

    A. *When* a tag is applied, select the `PayFail` tag.

    B. *Then* create a deal and choose the **Pay Fail** stage.

4. Following the instructions in *Chapter 7*, create an automation in the following sequence:

    A. *When* a deal enters the **Pay Fail** stage

    B. *Then* create a task.

    C. *Then* send an email. Add your three to five follow-up emails. These should have a sense of urgency warning of the consequences of nonpayment.

    D. Stop when a deal enters **Pay Success**.

5. Following the instructions in *Chapter 7*, create an automation in the following sequence:

    A. *When* a deal exits the **Pay Fail** stage

    B. *Then* remove a tag and select the `PayFail` tag.

6. Following the instructions in *Chapter 10*, add the `PayFail` tag to a widget on your dashboard.

How it works...

Managing payment failures using automation takes stress and uncomfortable conversations out of the equation while also ensuring that you're not leaving money on the table. By closely following failed invoices in your pipeline, you can effectively resend invoices and move deals from "failed" to "success" with ease.

For more information about suggested emails to use and a workflow map outlining the ideal steps for managing failed payments, check out the *bonus chapters* on GitHub: `https://github.com/PacktPublishing/Keap-Cookbook`

How essential sequences work together

Implementing essential automations is the very reason to have a CRM. It's all about saving precious time, effort, and—yes—even your hard-earned cash. By automating all those pesky repetitive tasks, you cut down on human errors and free yourself up to tackle big-ticket items. With these systems in place, you're not just efficient; you're consistent and scalable, too. That means you can get more done in less time, all while making the most of your resources and keeping costs down.

12

Data Management and Maintenance

When it comes to managing your CRM, one thing's for sure: keeping your list clean is non-negotiable. Think of it as the foundation of your CRM house – if it's shaky, everything else crumbles. Your reports? They're only as solid as the data they're built on. That's why regular and routine cleanup isn't just a suggestion; it's a must.

In this chapter, we'll dive deep into the world of hygiene basics. I'm talking about the nitty-gritty details – the stuff that keeps your CRM running smoothly day in, day out. So, buckle up and get ready to learn the ins and outs of list hygiene. Trust me, it's the key to unlocking the full potential of your CRM system.

In this chapter, we'll cover the following topics:

- Merging duplicate contact records
- Removing contacts who opt out
- Managing bounced emails

Technical requirements

Preparing to clean up your CRM is like getting your closet ready for a wardrobe makeover – exciting, daunting, but oh-so-rewarding in the end! Here are some steps to get you prepped and primed:

- **Assess your data**: Start by taking a good look around your CRM. Identify any duplicates, inaccuracies, or outdated information lurking in the depths. It's like sorting through your closet and tossing out those items you haven't worn in years. It's better to find potential issues on your own rather than having your client email you a 2 A.M. when they find a spelling error or wrong job title – and trust me, they will!

- **Have clear goals in mind**: What do you hope to achieve with your cleanup? Are you aiming to improve data accuracy, streamline processes, or enhance reporting capabilities? Setting clear goals will help guide your cleanup efforts and keep you focused along the way.

Merging duplicate contacts

Duplicate contacts can lead to conflicting or outdated information in your CRM. Imagine trying to reach out to a client, only to realize you've been working off of two different sets of contact details. It's a recipe for embarrassment and frustration.

How to do it...

Keap makes it easy to consolidate duplicate records.

Checking for duplicate records

Follow these steps:

1. Navigate to the **Settings** page and choose **Data Cleanup**:

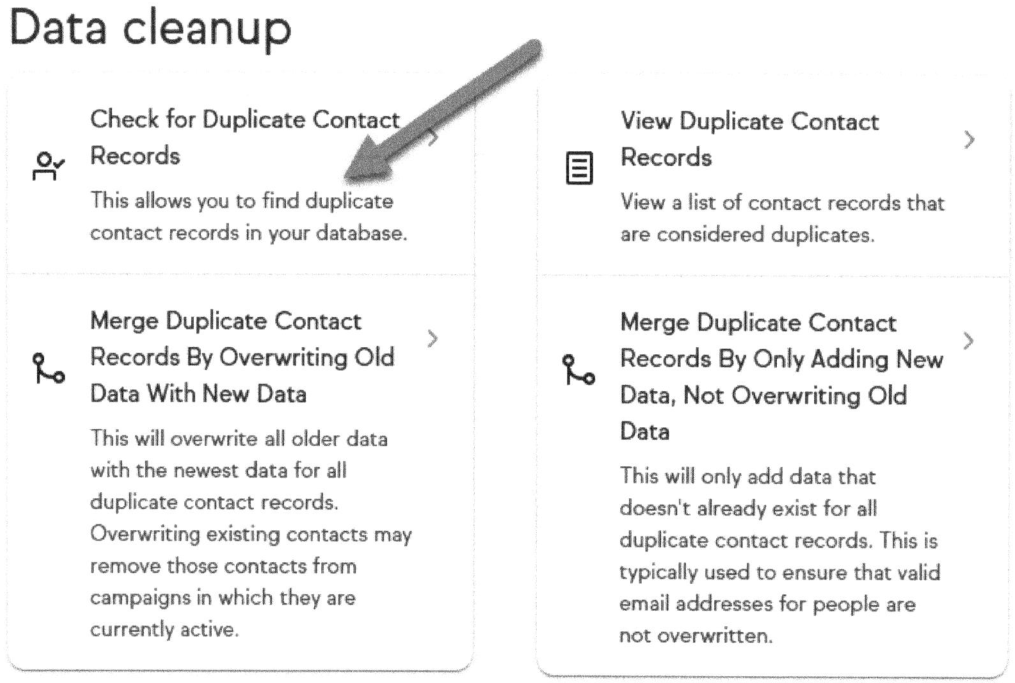

Figure 12.1 – Checking for duplicate contacts:

2. Keap provides a comprehensive explanation of duplicates. When you're done reviewing this material, click the **Next** button to continue.

3. Choose who to check:

    A. Check all records. (May take a long time).

    B. Check all records that haven't been marked as duplicates already.

    C. Check records that haven't been checked already.

4. For this recipe, choose the first option and click the **Next** button to continue.

5. Choose **Check Logic**. This is where you can decide how strict you want your criteria to be:

    I.   **One Field Check**: Only compares the email address

    II.   **Email** and one other field: Compares the email address and another field:

        i.   **Email** and **FirstName**.

        ii.   **Email** and **LastName**.

        iii. **Email** and **Fax1**.

        iv.   **Email** and **StreetAddress1**.

        v.   **Email** and **Phone1**.

    III. Using **FirstName**, **LastName**, and **One Other Field** compares multiple criteria:

        i.   **FirstName, LastName**, and **Email**.

        ii.   **FirstName, LastName**, and **Company**.

        iii. **FirstName, LastName**, and **StreetAddress1**.

        iv.   **FirstName, LastName**, and **Fax1**.

        v.   **FirstName, LastName**, and **Phone1**.

    For this recipe, we'll choose *option I*.

6. Let's choose **One Field Check** and click **Next** to continue.

7. In the confirmation pop-up, box click OK to start the deduping process.

8. When the process is complete, click the View the list of contact records that are considered duplicates link to review your potential duplicate records.

9. Compare the date of the original and duplicate record to ensure they truly are the same contact. If you're satisfied they're duplicates, click the [**Manual Merge**] link:

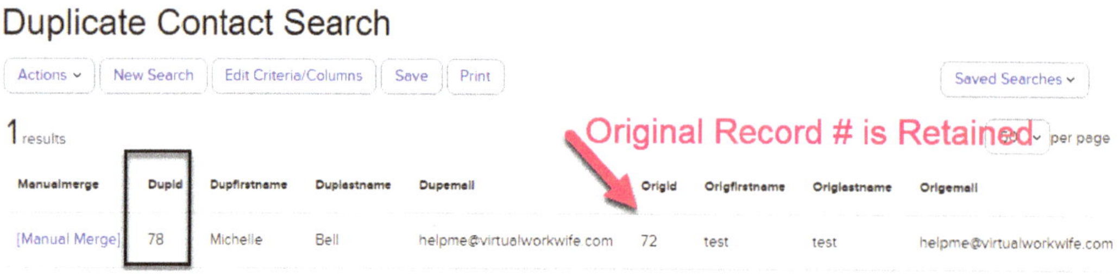

Duplicate Contact Search

Actions ⌄ | New Search | Edit Criteria/Columns | Save | Print | Saved Searches ⌄

1 results

Original Record # is Retained ⌄ per page

Manualmerge	DupId	Dupfirstname	Duplastname	Dupemail	OrigId	Origfirstname	Origlastname	Origemail
[Manual Merge]	78	Michelle	Bell	helpme@virtualworkwife.com	72	test	test	helpme@virtualworkwife.com

Figure 12.2 – Comparing duplicate records

10. You now have the option to manually choose which data to save to the consolidated contact. You can move data from the original contact or the more recent contact to the center column using the << and >> arrows. This will be the data that remains once the contacts are merged:

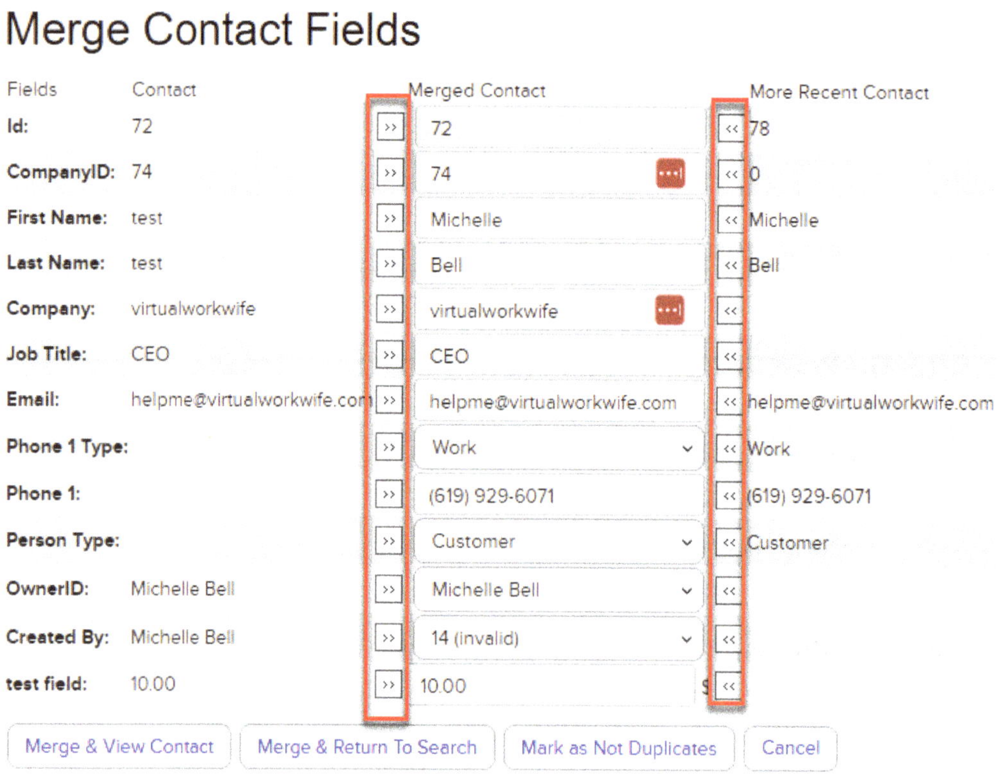

Merge Contact Fields

Fields	Contact		Merged Contact		More Recent Contact
Id:	72	>>	72	<<	78
CompanyID:	74	>>	74	<<	0
First Name:	test	>>	Michelle	<<	Michelle
Last Name:	test	>>	Bell	<<	Bell
Company:	virtualworkwife	>>	virtualworkwife	<<	
Job Title:	CEO	>>	CEO	<<	
Email:	helpme@virtualworkwife.com	>>	helpme@virtualworkwife.com	<<	helpme@virtualworkwife.com
Phone 1 Type:		>>	Work	<<	Work
Phone 1:		>>	(619) 929-6071	<<	(619) 929-6071
Person Type:		>>	Customer	<<	Customer
OwnerID:	Michelle Bell	>>	Michelle Bell	<<	
Created By:	Michelle Bell	>>	14 (invalid)	<<	
test field:	10.00	>>	10.00	<<	

Merge & View Contact | Merge & Return To Search | Mark as Not Duplicates | Cancel

Figure 12.3 – Choosing what data to keep

11. Choose one of the following options to continue:

   A. **Merge & View Contact**: Merges the data and opens the contract record.

   B. **Merge & Return to Search**: Merges the data and returns you to the previous screen.

   C. **Mark as Not Duplicates**: There are occasions when what seems like duplicate contacts are two separate relationships. Take a husband and wife who share an email address as an example. In this case, choosing to mark them as not duplicates will remove them from your cleanup list.

   D. **Cancel**: Returns you to the previous screen.

How it works...

Overall, regularly cleaning up duplicate contacts helps you maintain the integrity, efficiency, and effectiveness of your CRM system, ultimately leading to better reporting, business outcomes, and improved customer relationships.

Managing opt-outs

Managing opt-outs, also known as unsubscribes, is a crucial aspect of email marketing compliance and CRM. It involves handling requests from recipients who no longer wish to receive emails from your mailing list.

How to do it...

Keap automatically tracks various interactions with contacts, including email opens, link clicks, and opt-outs.

Identifying opt-outs

Follow these steps:

1. Click on the **REPORTS** tab in the left-hand side navigation bar.

2. Under the **Contact tracker** header select **Email engagement tracker**.

3. Click the **Reset Filters** button to start a new search.

4. Choose the following criteria:

A. In the **Email Status** dropdown, select **Contains Any**.

B. Click inside the search box and select all the opt-out options, as shown in *Figure 12.4*:

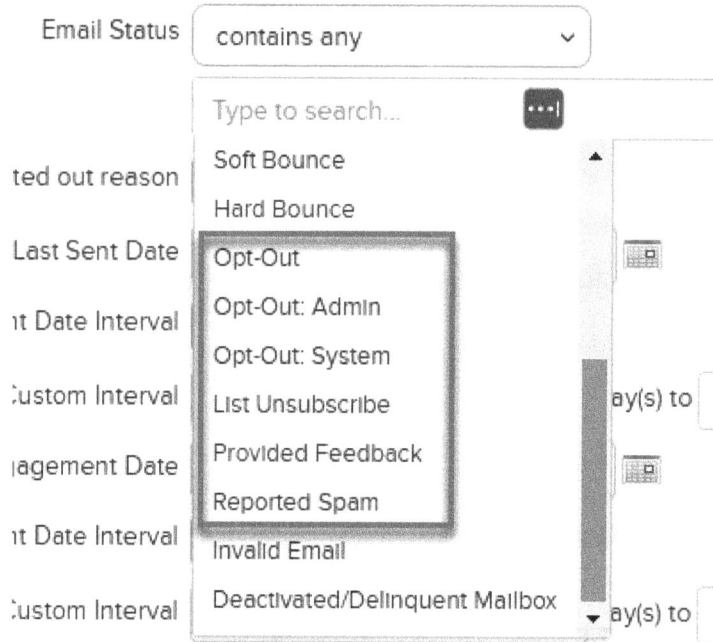

Figure 12.4 – Selecting Opt-Out as a criterion

It's important to note that while recipients have the right to unsubscribe from marketing emails, businesses may still need to retain certain information about them for legal or operational purposes. It's essential to balance compliance with data protection regulations and the need to maintain accurate records:

1. To ensure you retain proper documentation, do the following:

I. Click on the **Misc Criteria** tab.

II. In the **Purchased Products** dropdown, select **Doesn't have ANY of these**.

III. Click inside the search box and select all your products. Hold down Shift + A or click the first product and hold Shift, then click the bottom product.

2. Click **Search** to view your results.

You will now be presented with a list of opt-outs who have not made purchases. It is safe to remove these people from your CRM.

Deleting opt-outs

Follow these steps:

> **Note**
>
> It is highly recommended that you export your list of opt-outs *before* you delete them as a backup.

1. Check the box next to each contact you want to remove or use the **check all** box at the top of the list.
2. Click the **Actions** button.
3. Select **Export Contacts**.
4. By default, every data field is chosen.
5. Click the **Export** button.
6. A pop-up box will appear, confirming that your download is in progress. Click the **Ok, got it** button to close it.
7. Click the **Cancel** button to return to your search.
8. Click the **Actions** button again, this time choosing **Delete Contacts**.
9. A warning message will appear, confirming the number of contacts you are about to delete:

    A. Check the **Yes, I want to delete all xx records in the filter!** button to proceed.

    B. Alternatively, click the **Cancel** button to return to your search.

10. To proceed, click the **Process Action** button.

Maintaining opt-outs with purchase history

While you may not want to delete opt-outs that have made purchases, you also don't want to include them in any future broadcast emails. One way to control this would be to tag them with **Do Not Email**. Doing so will allow you to use the tag to exclude people who can't receive emails from your broadcasts. Follow these steps:

1. Click the **Reset Filters** button to start a new search.
2. Choose the following criteria:

    A. In the **Email Status** dropdown, select **contains any**.

    B. Click inside the search box and select all the opt-out options, as shown in *Figure 12.4*.

3. Click on the **Misc Criteria** tab.
4. In the **Purchased Products** dropdown, select **Has ANY of these**.

5.   Click inside the search box and select all your products. Hold Shift + A or click the first product and hold Shift then click the bottom product.

6.   Click **Search** to view your results.

You will be presented with a list of opt-outs who have made purchases. Next, we want to apply the proper tag.

1.   Check the box next to each contact you want to tag or use the **check all** box at the top of the list.

2.   Click the **Actions** button.

3.   Choose **Apply/Remove Tag**:

   I.   Be sure to click the **Apply** option.

   II.   In the **Apply These Tags** box, select your tag.

   III. Alternatively, click the **Create a new tag** link to open the creation box:

      i.   Give your tag a name, such as Do **Not Email**.

      ii.   Choose a category or type a category name in the provided space.

      iii. Click the **Create this Tag** button.

4.   Click **Process Action** to continue:

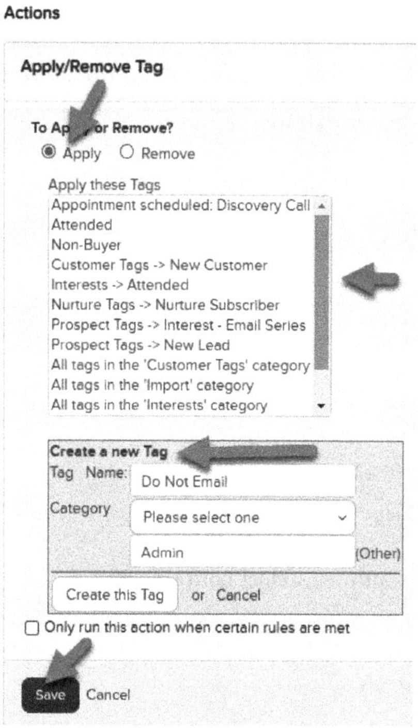

Figure 12.5 – Choosing or creating a tag

How it works...

Overall, managing opt-outs effectively is essential for maintaining a positive reputation, complying with regulations, and fostering trust and transparency with your email subscribers. By implementing clear opt-out mechanisms and promptly honoring unsubscribe requests, you can demonstrate your commitment to respecting recipients' preferences and building strong customer relationships.

By default, Keap will not send an email to an opted-out address, even if you've included it in your broadcast or automation sequence. While it might seem harmless to leave them on your list, it does effectively water down your tracking and reporting results. It boils down to the adage "crap in, crap out…" If you're feeding your report tools bad information, you're going to get bad reports!

Managing bounced emails

Managing bounced and invalid emails in your CRM is essential for maintaining data accuracy, improving email deliverability, and preserving the sender's reputation. Bounced and invalid emails can clutter your CRM with inaccurate or outdated information. By regularly identifying and removing these entries, you ensure that your database remains clean, up-to-date, and reliable for effective communication and decision-making.

How to do it...

Managing bounced and invalid emails in Keap is easy if you follow these steps,

Identifying bounced email addresses

Follow these steps:

1. Click on the **REPORTS** tab in the left-hand side navigation bar.
2. Under the **Contact tracker** header, select **Email engagement tracker**.
3. Click the **Reset Filters** button to start a new search.
4. Choose the following criteria:

    A. In the **Email Status** dropdown, select **contains any**.
    B. Click inside the search box and select **Hard Bounce**.

As we discussed in the previous recipe, you may still need to retain certain information about contacts for legal or operational purposes. To do so, follow these steps:

1. Click on the **Misc Criteria** tab:

    A. In the **Purchased Products** dropdown, select **Doesn't have ANY of these**.

    B. Click inside the search box and select all your products. Hold down *Shift + A* or click the first product and hold *Shift* then click the bottom product.

2. Click **Search** to view your results.

You will now be presented with a list of hard bounces who have not made purchases. It's safe to remove these people from your CRM.

Deleting bounced emails

Follow these steps:

> **Note**
>
> It is highly recommended that you export your list of opt-outs *before* you delete them as a backup.

1. Check the box next to each contact you want to remove or use the **check all** box at the top of the list.

2. Click the **Actions** button

3. Select **Export Contacts**.

4. By default, every data field is chosen.

5. Click the **Export** button:

    A. A pop-up box will appear to confirm that your download is in progress. Click the **OK, got it** button to close this.

    B. Alternatively, click the **Cancel** button to return to your search.

6. Click the **Actions** button again, this time choosing **Delete Contacts**.

7. A warning message will appear, confirming the number of contacts you are about to delete:

    A. Check the **Yes, I want to delete all xx records in the filter!** box to proceed.

    B. Alternatively, click the **Cancel** button to return to your search.

8. To proceed, click the **Process Action** button.

Maintaining hard bounces with purchase history

While you may not want to delete opt-outs that have made purchases, you also don't want to include them in any future broadcast emails. One way to control this would be to tag them with **Do Not Email**. Doing so will allow you to use the tag to exclude people who can't receive emails from your broadcasts. Follow these steps:

1. Click the **Reset Filters** button to start a new search.

2. Choose the following criteria:

    A. In the **Email Status** dropdown, select **contains any**.

    B. Click inside the search box and select **Hard Bounce**.

3. Click on the **Misc Criteria** tab.

4. In the **Purchased Products** dropdown, select **has ANY of these**.

5. Click inside the search box and select all your products. Hold down Shift + A or click the first product and hold Shift then click the bottom product.

6. Click **Search** to view your results

You will now be presented with a list of hard bounces who have made purchases. Next, we want to apply the proper tag. Follow these steps:

1. Check the box next to each contact you want to tag or use the **check all** box at the top of the list.

2. Click the **Actions** button

3. Choose **Apply/Remove Tag**:

    I. Be sure to click the **Apply** option.

    II. In the **Apply These Tags** box, select your tag.

    III. Alternatively, click the **Create a new tag** link to open the creation box:

        i. Give your tag a name.

        ii. Choose a category or type a category name in the provided space.

        iii. Click the **Create Tag** button.

4. Click **Process Action** to continue.

How it works...

Overall, managing bounced and invalid emails in your CRM is essential for optimizing email deliverability, preserving sender reputation, ensuring compliance, and fostering meaningful engagement with your audience. By prioritizing data cleanliness and accuracy, you can maximize the effectiveness of your email marketing efforts and build stronger relationships with your customers.

Fixing or deleting invalid emails

Making sure your email addresses stored in Keap are accurate is crucial for maintaining effective communication and data integrity. Fixing those annoying invalid addresses doesn't just reduce bounces; it also enhances your email deliverability, boosting your credibility in the process. But it's not just about smooth sailing; it's about being savvy with your resources too. Cleaning up those invalid emails not only saves on storage costs but also ensures your resources are utilized efficiently. And when you maintain accuracy with valid email addresses, you're not just following the rules; you're building trust with your audience. Adhering to regulations such as the CAN-SPAM Act and GDPR demonstrates your commitment to privacy and earns respect from your subscribers.

How to do it...

Fixing invalid emails is somewhat hit-and-miss. When someone types `.cmo` instead of `.com`, you can easily fix it and get them back into your marketing stream. Other issues may not be so simple to deduce. It may require doing some research or, gasp, actually calling someone!

Identifying invalid email addresses

Follow these steps:

1. Click on the **Reports** tab in the left-hand side navigation bar.
2. Under the contact tracker header, select **Email engagement tracker**.
3. Click the **Reset Filters** button to start a new search.
4. Choose the following criteria:

    A. In the **Email Status** dropdown, select **contains any**.

    B. Click inside the search box and select **Invalid Email**.

5. Click on the **Misc Criteria** tab.

    A. In the **Purchased Products** dropdown, select **Doesn't have ANY of these**.

    B. click inside the search box and select all your products. Hold down *Shift + A* or click the first product and hold *Shift* then click the bottom product.

6. Click **Search** to view your results.

You will be presented with a list of invalid emails who have not made purchases.

> **Note**
> At the time of writing, returning to your search results once you begin editing is not a smooth process. You can save yourself a lot of time and effort in this next step by opening the record in a new tab.

Editing or deleting an email address

Follow these steps:

1. Click on the **Name** value to open the contact record (PC users must right-click and choose **Open in a new tab**).

2. In the top-right corner, click **Info**.

3. Click the **Edit** button.

4. You can now do one of the following things:

   A. Edit the address and click **Save**.

   B. Scroll to the bottom of the page and click the **Delete contact** button.

How it works...

The sooner you can pinpoint and resolve broken or invalid emails, the greater your odds of turning them from leads into buyers. Remember, each email you acquire represents an investment of your time and resources. That's why it's vital to stay on top of list maintenance and address any issues before they escalate. By staying proactive, you can ensure that your email marketing efforts are as efficient and effective as possible, leading to greater success in converting leads into customers.

Index

A

advanced automation
building 193-199
advanced automation builder
canvas, navigating 173
connection, deleting 176
elements, connecting 175, 176
elements, deleting 174
elements, renaming 175
keyboard shortcuts 173
multiple goals or sequences,
moving as group 174
using 172
alternative text 146
appointment timer 183
appointment types
setting up 29-32
working 32, 33
automation 176
automation version history
accessing 176, 177
Make a Copy option 178
restoring 176
Revert Changes option 177

B

bounced emails
addresses, identifying 263, 264
deleting 264
hard bounces, maintaining with
purchase history 265
managing 263
working 265
business profile
setting 24, 25
working 26

C

calendar connection 26, 27
working 28
CAN-SPAM compliance 24
checkout forms
creating 117, 118
follow-ups, automating for 123, 124
promo code, adding 120-122
thank you page, adding 124
upsell, adding 119, 120
companies
adding 45-47

contact record

  quote, creating from 106

Contact Relationship Management (CRM) 3

contacts

  adding manually 42, 44

  grouping 47-49

  importing 50-52

  searching 60, 61

contact tracker reports

  generating 209

  saved contact tracker report, viewing 211

  saving 210

  viewing 209, 210

  working with 208

Contact Us form 247, 248

  working 249

content, landing page 143

  button, adding 147, 148

  columns 144

  columns, adding 144

  divider, adding 148, 149

  heading 144

  images, adding 146, 147

  text 145

  text, adding 145

customer relationship
management (CRM) 15

custom fields

  adding 52-55

D

daily routine

  dashboard widgets, displaying 238

  dashboard widgets, hiding 238

  group widgets, adding 239, 240

  managing 237, 238

  widgets, arranging on dashboard 241

date timer 182

deal

  creating, manually 227, 228

  moving backward 229

  moving forward manually 228

  pipeline, changing 230

  pipeline, deleting 230

decision diamonds

  configuring 189, 190

  rules, copying 192

  rules, deleting 192

  rules, importing from another sequence 191

  working 192

  working with 188, 189

delay timer 181

discovery call 249, 250

  working 251

documents

  managing 216

Domain-based Message
Authentication, Reporting, &
Conformance (DMARC) 68

double booking 26

duplicate contacts

  duplicate records, checking 256-259

  merging 256

  working 259

E

Easy Automations

  contacts, viewing and removing in 167

  creating 162-165

  deactivating 167

  deleting 168

  editing 166

  framework 168

  prebuilt automation template, using 166

reports, checking 168

stop rule, adding to 165, 166

working with 162

Easy Automations, framework

stop conditions 170

then conditions 169

when condition 168, 169

email broadcast

reports, reviewing 96-98

email builder 66-70

email connection 28

working 29

email service provider (ESP) 98

email template

blocks, attaching to 86

body 86, 87

content, creating 72, 73

creating 71

Images section, adding 87

merge fields 88, 89

sections 71

sections, testing and sending 89-91

Uploads section 88

email template, content

appointments 84, 85

buttons 77, 78

columns 74, 75

dividers 78, 79

files 85, 86

heading 75

HTML 79

images 76, 77

menu 79, 80

signatures 83

social icons 81

text 75, 76

timers 81-83

videos 83

essential sequences

working 254

F

field timer 182

field types

date fields 53

option lists 53

specially formatted fields 54

text and number fields 53

files

managing 216, 217

uploaded file, deleting 218

uploaded file, sending to contact 219, 220

uploaded files, renaming 218

uploaded files, viewing 217

uploading, to contact record 217

G

goals 178

triggered, by contact 178, 179

triggered, by Keap user or automation 179

Google Reviews 38

requesting 235, 237

working 39

H

HTML, landing page

adding 149

appointment, adding 154

Checkout form 154

forms, adding 153, 154

menu, adding 149, 150

social icons, adding 150, 151

timer, adding 151, 152

video, adding 152

I

internal forms 136

automating 138-140

building 136

existing contacts, updating within 141, 142

fields, adding 137

fields, editing 138

fields, removing 138

publishing 140

used, for adding contacts 141

invalid email addresses

deleting 267

editing 267

identifying 266

invalid emails 266

invoices

creating 112, 113

deposits, applying 114

discounts, adding 114

generating 111, 112

More actions dropdown 117

notes and terms, adding 115, 116

online payment, accepting 115

reviewing and sending 116

sales tax, adding 113, 114

K

Keap account

application name 5

Keap dashboard 11

using 11, 12

Keap Max 4

Keap mobile app

initial setup 12

multiple account, adding 13

setting up 12

Touch ID or Face ID, used for logging in 14

Keap Partner 23

Keap phone numbers

marketing number, selecting 17, 18

obtaining 16, 17

obtaining, for business line 17

working 19, 20

Keap Pro 4

Keap Pro/Max

navigation system 6-10

Keap Ultimate 5

Keap version

identifying 5

L

landing page 142

audit 156, 157

blocks 155

body 155

content 143

HTML 149

images 156

pages 157

publishing 157, 158

settings 157

uploads 156

lead magnet 251-253

M

manual automations 223

contact, adding to advanced automation 224, 225

contact, removing from Easy Automation 223, 224

process, simplifying 225, 226

manual payments
adding 220, 221
invoices and quotes, sending 222
posting 220
working 223
merchant account 33
My day page
appointments, managing 241
tasks, managing 242, 243
utilizing 241

N

newsletter opt-in 246
working 247
notes
adding, to contact 231
deleting 232
editing 232
using 230

O

opt-outs
deleting 261
identifying 259, 260
maintaining, with purchase history 261
managing 259
working 263

P

pay (payment) fail
managing 254
requisites 253, 254
payment processing 33-35
personal avatar
creating 20, 21
working 21

products
creating 35-37
public forms 128
automating 133, 134
building 129
deleting 135
fields, adding 130
fields, editing 131
fields, removing 131
prerequisites, for creating
web forms 128, 129
publishing 134, 135
styling 131, 132

Q

quotes
creating 104, 105
creating, from contact record 106
creating, from Sales menu 105, 106
customizing 106, 108
files, attaching to 108
More actions dropdown 110, 111
sending, to recipient 109, 110
using, in sales process 104

R

Reset Filters button 206

S

Sales menu
quote, creating from 105, 106
sales pipeline
creating 99-104
deal, adding to 102, 103
deal, moving to difference pipeline 103, 104

sales reports

All sales report widget, using 202, 203

generating 203, 204

saved report, viewing 208

saving 207

viewing 204-207

working with 202, 203

sequences 180

communications 183

processes 183-188

timers 181-183

start timer 181

T

tagging 56

tags

applying 59

deleting 58

editing 58

removing 59

working with 57

tasks

adding, to contact 233

deleting 234

editing 234

marking, as complete 234

using 232, 233

templates, for text broadcasting

creating 94-96

text message broadcasts 91-94

timers 181

Appointment Timer 183

date timer 182

delay timer 181

field timer 182

start timer 181

U

unsubscribes 259

uploaded files

deleting 218

renaming 218

sending, to contact 219

viewing 218

user accounts

managing 21, 22

users, adding in Keap 22, 23

users, deactivating in Keap 23

user types 24

user types 24

V

versions

Keap Max 4

Keap Pro 4

Keap Ultimate 5

Z

Zapier integration 37

working 37

‹packt›

packtpub.com

Subscribe to our online digital library for full access to over 7,000 books and videos, as well as industry leading tools to help you plan your personal development and advance your career. For more information, please visit our website.

Why subscribe?

- Spend less time learning and more time coding with practical eBooks and Videos from over 4,000 industry professionals
- Improve your learning with Skill Plans built especially for you
- Get a free eBook or video every month
- Fully searchable for easy access to vital information
- Copy and paste, print, and bookmark content

Did you know that Packt offers eBook versions of every book published, with PDF and ePub files available? You can upgrade to the eBook version at packtpub.com and as a print book customer, you are entitled to a discount on the eBook copy. Get in touch with us at customercare@packtpub.com for more details.

At www.packtpub.com, you can also read a collection of free technical articles, sign up for a range of free newsletters, and receive exclusive discounts and offers on Packt books and eBooks.

Other Books You May Enjoy

If you enjoyed this book, you may be interested in these other books by Packt:

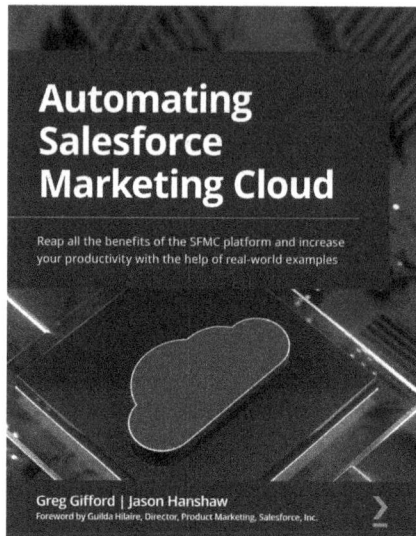

Automating Salesforce Marketing Cloud

Greg Gifford, Jason Hanshaw

ISBN: 978-1-80323-719-0

- Understand automation to make the most of the SFMC platform

- Optimize ETL activities, data import integrations, data segmentations, email sends, and more

- Explore different ways to use scripting and API calls to increase Automation Studio efficiency

- Identify opportunities for automation with custom integrations and third-party solutions

- Optimize usage of SFMC by building on the core concepts of custom integrations and third-party tools

- Maximize utilization of employee skills and capabilities and reduce operational costs while increasing output

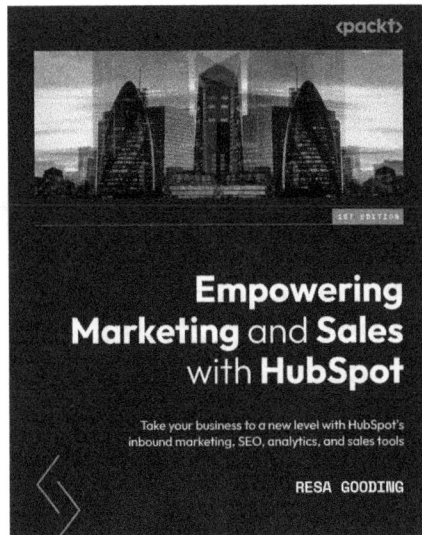

Empowering Marketing and Sales with HubSpot

Resa Gooding

ISBN: 978-1-83898-714-5

- Explore essential steps involved in implementing HubSpot correctly
- Build ideal marketing and sales campaigns for your organization
- Manage your sales process and empower your sales teams using HubSpot
- Get buy-in from your management and colleagues by setting up useful reports
- Use Flywheel strategies to increase sales for your business
- Apply the inbound methodology to scale your marketing
- Re-engage your existing database using the HubSpot retargeting ads tool
- Understand how to use HubSpot for any B2B industry in which you operate

Packt is searching for authors like you

If you're interested in becoming an author for Packt, please visit `authors.packtpub.com` and apply today. We have worked with thousands of developers and tech professionals, just like you, to help them share their insight with the global tech community. You can make a general application, apply for a specific hot topic that we are recruiting an author for, or submit your own idea.

Share Your Thoughts

Now you've finished *Keap Cookbook*, we'd love to hear your thoughts! Scan the QR code below to go straight to the Amazon review page for this book and share your feedback or leave a review on the site that you purchased it from.

https://packt.link/r/1-805-12949-X

Your review is important to us and the tech community and will help us make sure we're delivering excellent quality content.

Download a free PDF copy of this book

Thanks for purchasing this book!

Do you like to read on the go but are unable to carry your print books everywhere?

Is your eBook purchase not compatible with the device of your choice?

Don't worry, now with every Packt book you get a DRM-free PDF version of that book at no cost.

Read anywhere, any place, on any device. Search, copy, and paste code from your favorite technical books directly into your application.

The perks don't stop there, you can get exclusive access to discounts, newsletters, and great free content in your inbox daily

Follow these simple steps to get the benefits:

1. Scan the QR code or visit the link below

https://packt.link/free-ebook/9781805129493

2. Submit your proof of purchase
3. That's it! We'll send your free PDF and other benefits to your email directly

www.ingramcontent.com/pod-product-compliance
Lightning Source LLC
Chambersburg PA
CBHW061805210326
41599CB00034B/6880

* 9 7 8 1 8 0 5 1 2 9 4 9 3 *